I0005996

DISCLAIMER

Printed in U.S.A.

Available from

National Technical Information Service
U.S. Department of Commerce
5285 Port Royal Road
Springfield, VA 22161

CONTENTS

Hypercube Algorithms and Implementations [1]

Oliver A. McBryan [2,3]

Eric F. Van de Velde [2]

Courant Institute of Mathematical Sciences,
New York University
New York, N.Y. 10012

ABSTRACT

Parallel algorithms are presented for important components of Computational Fluid Dynamics algorithms along with implementations on hypercube computers. Elliptic equations with 1 6 million unknowns and hyperbolic equations with 4 million unknowns have been solved using 128 processors

For elliptic equations, a parallel Preconditioned Conjugate Gradient method is described which has been used to solve pressure equations discretized with high-order finite elements on irregular grids A parallel Full Multigrid Method and a parallel Fast Poisson Solver are also presented Hyperbolic Conservation Laws have been discretized with parallel versions of finite difference methods and with the Random Choice Method

The performance of these algorithms is analyzed in terms of machine efficiency, communication time, bottlenecks and software development costs A key aspect of this work is the development of a library of parallel operators for distributed vectors and matrices, efficient for both full and sparse data The implementation of these operators on hypercubes is described along with measurements of communication effects Using the library the PDE algorithms mentioned above have been implemented on both serial computers and on hypercubes without *any* code modification All inter-process communication is hidden in library routines A general parallel computer simulator is described along with its use in the development of the algorithms

The relation of the model problems solved here to the more complex physical problems encountered in real fluids is discussed Techniques are developed for comparing the behavior of an algorithm on different architectures as a function of problem size and local computational effort

---

1 Presented to the 2nd SIAM Conference on Parallel Computing, Norfolk, Nov 18 21, 1985
2 Supported in part by DOE contract DE-ACO2-76ER03077
3 Supported in part by NSF grant DMS 83 12229

Hypercube Algorithms and Implementations

Oliver A. McBryan

Eric F. Van de Velde

Courant Institute of Mathematical Sciences,
New York University

1. INTRODUCTION

In this paper we describe the implementation of partial differential equation solvers on hypercube processors, part of an ongoing effort to exploit parallelism in the solution of equations arising in Computational Fluid Dynamics. We present results of computations run on the Caltech Mark II Hypercube parallel processor and on the Intel iPSC d7 processor. These machines have respectively 32 and 128 processors connected on 5 and 7 dimensional hypercube networks. We have previously described the implementation of some of these algorithms on the shared memory Denelcor HEP parallel computer.[1,2,3,4,5,6,7,8]

Our goal is to develop algorithms that make close to optimal use of available parallelism on a range of realistic problems. Our studies have concerned the development of parallel algorithms for elliptic, parabolic and hyperbolic equations. These algorithms include important components of real fluid dynamics codes. All of our algorithms are for equations with two space dimensions, though extension to three dimensions is straight-forward in each case. While most of the computational results described in the paper relate to PDE solution, the techniques used are much more general. The PDE solvers are written entirely in terms of a general-purpose library of parallel operations which automates the construction of distributed data structures and provides parallel operations on them. Much of the paper (sections 3-7) will concentrate on this library, describing its implementation on hypercubes, and the techniques used to make it efficient for such architectures. In section 1.1 we give an overview of this library, followed in section 1.2 by an introduction to the PDE problems we have solved. Section 1.3 discusses the relation of the model problems solved in this paper to parallelization of complex physical problems and introduces a method for performance analysis that provides useful comparison of parallelization efficiency between different algorithms and machines. In section 1.4 we make some comments relevant to comparison of our studies on the Caltech Hypercube and on the Intel ıPSC Section 1.5 surveys the remainder of the paper ın more detaıl

1.1. Portable Parallel Software and Libraries

A primary long-range aim of our work is the development of portable programming methodologıes for numerıcal applıcations The goal ıs to be able to use the same code on a wide range of dıfferent processors, includıng serial, SIMD and MIMD machines, wıth both shared memory and ınessage-passing architectures We report progress ın thıs area. In partıcular, we have ımplemented a lıbrary of vector and matrıx operations for hypercubes Thıs library deals with globally distributed vectors and matrices. It ıncludes allo-catıon routines for such objects, standard linear algebra operators such as ınner products, vector maximum, vector sums and products, matrıx transpose and matrıx-vector multıply. The allocation routines allow vectors and matrices to be dıstrıbuted ın various ways across the processors, and routines are provıded to convert between dıfferent representatıons. The lıbrary ıs effıcıent for both sparse matrices such as those encountered ın dıscretızıng partıal dıfferentıal equatıons and for full matrices A second library ınvolves routines for object-level communication, both between hypercube nodes as well as with the host processor. All of our applications are implemented in terms of these two libraries.

There are several advantages that result from the use of these lıbraries, beyond the normal improvement ın software modularıty Some of the lıbrary routınes ınvolve substantıal and complex ınter-process communicatıon on the hypercube - for example the *inner_product*(), *transpose*() and *matrıx_vector*() routines for dıstrıbuted vectors and matrıces. These library routines may be written for different parallel processors as well as for serıal and vector pro-cessors and represent a powerful mechanısm for hıdıng communıcatıon and network dependence in programs As a result, most of our applicatıons algo-rıthms can be run on both serıal machines and hypercubes wıth *no* code modification In particular we ıdentify a class of algorıthms, which we term *Transpose-Splıt Algorıthms* whıch have this property We present a general dıscussion of such algorithms as well as several applications to hyperbolic and ellıptic problems

We have also observed that a large number of algorithms, when written in terms of our library, take the form of SIMD rather than MIMD programs In particular the code running in each node is identical and does not refer in any way to location in the network (of course such dependencies are buried in the library routines) Furthermore we do not use cooperating host programs which only tend to complicate the clarity of algorithms, and introduce critical sections into programs We believe that by casting programs in an SIMD form, portability is greatly enhanced, and programs can be easier to debug and understand.

A critical issue here is the semantics of the vector library routines. We illustrate this with the example of the inner product of two distributed vectors In our approach, the inner product operation is called simultaneously by all processors, and all return the (same) global value. We do not allow the possibility that a single processor compute a global inner product. Consequently all processors compute quantities such as inner products that are used to determine when to terminate an iteration Each processor independently decides to terminate its iteration, although of course all terminate at the same point since the same inner product value is obtained at all nodes An alternate approach would be to assign one processor (or the host) as a control processor to monitor convergence On detecting sufficient accuracy that processor would then notify all others to terminate. This scheme does not have an SIMD form since one processor is distinguished There is also no real time savings in this scheme because although most processors avoid computing the inner product, they have to wait while the control processor computes the inner product and notifies them to continue iterating or to terminate.

1.2. Elliptic and Hyperbolic Equations

Once the parallel libraries are available, implementation of various solution methods for PDE becomes straightforward All of the basic parallel data-structures required for storing data across the processors are automatically generated by calls to library routines Further library routines are available to perform all operations on these data structures that require any communication. The remaining code is then related to only one processor, and is just standard serial code - we have not developed any new numerical methods for solving PDE. The algorithms we have used include random choice and finite difference methods for hyperbolic equations, multigrid methods for Poisson-type equations and preconditioned conjugate gradient methods for arbitrary-order finite element discretizations of elliptic equations on possibly irregular grids

The hyperbolic equations are the easiest to parallelize, and in fact we have parallelized various hyperbolic solvers by adding as little as one line of code to the corresponding serial code - see section 9 3 below. We have used several parallelization strategies for hyperbolic equations and we compare these from the point of view of computational efficiency. The simplest method is not the most efficient However it is quite sufficient as computation already substantially dominates communication costs even for this method From a software point of view we prefer the simpler approach - trading a small amount of efficiency for a large decrease in software complexity. Furthermore we have used the same parallelization strategy, which we call *Transpose Splitting Parallelization*, for various other applications, including an FFT-based fast Poisson solver - see sections 8 and 14. The basic observation underlying this approach is that an efficient parallel matrix transpose can be used to parallelize a numerical algorithm of ADI or operator splitting type.

Another general parallelization technique, which we call *Areal Decomposition* is applicable to both hyperbolic equations and to multigrid methods and other relaxation schemes Here a grid is decomposed into rectangular sub-grids which are distributed in an areal (or volume) manner to a set of processors Extra boundary layers are provided around each sub-grid and are used to store duplicate data recording values at grid-points in neighboring processors out to a specified distance, typically 1 grid-point A basic library routine is used to globally allocate such data structures and another routine (*shuffle*) can be called to update the boundary duplicates at any time The combination of the data structure and the *shuffle* operation simulates a shared memory extremely closely - in fact does simulate one exactly for grid operators whose stencil range does not exceed the buffer range Efficiency depends on the aspect ratio of the sub-grids used and we measure this area-perimeter effect in several applications.

1.3. Real Problems and Performance Comparisons

For the most part we have studied model problems which have simpler governing equations than those for realistic physics. We have studied the issue of how effectively real problems may be parallelized. Real physical equations differ from our model equations in complexity For example expensive equations ot state may need to be solved at every grid-point Because of the resulting increase in computation time per grid-point, with little or no increase in communication time, computational efficiency will in general oe higher for many realistic problems than for our model problems. Since the model problems already behave extremely well, the prognosis for the real problems looks excellent.

We have studied this question using a model hyperbolic code that does nothing but compute *NFLOPS* floating point operations per grid point and performs the standard inter-process communication typical of hyperbolic solvers We then study the computational efficiency of this code as a function of grid-size, number of processors and *NFLOPS* The conclusion is that attaining high efficiency is extremely easy even on very coarse grids - in fact efficiencies over 90% are attainable on the Intel iPSC with *NFLOPS* of under 100, on grids so coarse that there are only a few dozen grid-points per processor, see section 10 for details Caltech Hypercube efficiencies are generally even higher We obtain efficiency performance curves for various grid sizes that essentially characterize the behavior of the hypercube in question for a very wide range of numerical processes Given an estimate of *NFLOPS* we can predict performance of an algorithm on various grids or with increasing numbers of processors. Furthermore comparing such curves for different algorithms gives an immediate comparison of the algorithms over the range of parameters Plotting equivalent curves for different processors would provide a useful way to compare hardware designs over a range of different problems.

Variations in hardware design can play a major role in the behavior of algorithms, even among machines with the same network configuration Critical parameters include the amount of memory per processor, the network communication rate, the overhead for sending short messages and the overall balance between these quantities and the processor speed This has shown up very clearly in our work with the Caltech Hypercube and the Intel iPSC Despite the larger memory and somewhat faster processing speed of the iPSC, the Caltech machine is more successful for most algorithms because there is almost no overhead for short messages, whereas with the iPSC all messages up to 1024 bytes in length take essentially the same time. The question of communication startup cost is closely tied to the available memory. In the current iPSC design, the memory available (about 250 Kbytes per node) is such that for many algorithms, communication is just beginning to be efficient even in the largest problems that can fit on the processors. For example, if a square subgrid of maximal size is stored on an iPSC processor, it will contain about 2^{16} real-valued grid-points and consequently have edges containing about 256 grid-points (1024 bytes of data) A communication in which a whole edge is communicated at once to another processor is then barely in the efficient range In practice, memory is required for program storage and other data so that the situation is actually less favorable The Caltech Hypercube does not suffer from this problem In fact on that machine even individual boundary data points of a subgrid may be efficiently transferred to a neighboring processor. It follows that such *areal decompositions* of grids are substantially more favorable on the Caltech machine.

In analyzing algorithms in the sequel we will comment where appropriate on these issues. Such considerations play an especially important role in the asymptotic analysis of algorithms. An otherwise accurate analysis may be seriously flawed in practice if communication startup overhead is omitted In section 2 3 we introduce three critical parameters α, β and γ which characterize the communication and computation performance of parallel computers These quantities are used repeatedly throughout the paper.

1.4. Caltech Hypercube vs. Intel iPSC

We have referred above to advantages of the Caltech design over the Intel iPSC. The iPSC has advantages too, such as more memory and processors The results of our experiments performed on the Caltech and Intel hypercubes are not for the most part directly comparable We describe now some of the reasons for this

We performed all of the Caltech Hypercube experiments prior to any of the iPSC experiments The former experiments were performed with generally far inferior algorithms to the latter In part we had learned from the Caltech experiments, and in part we were forced to develop better algorithms by the difficulties of dealing with the iPSC communications overhead As an example, when we first moved a multigrid code from the Caltech Hypercube to the Intel iPSC it ran 100 times *slower* on the latter machine! Analysis showed that control messages being sent between the host computer and the node processors were dominating all computation The fundamental problem was that control messages were typically a few words long and were therefore hundreds of times less efficient than on the Caltech cube. Another problem appeared to be that the host is time-shared and may not be ready to interrupt itself immediately on receipt of a message from the cube, effectively increasing communication costs with nodes. A third problem was that boundary data on the edges of subgrids were transferred to neighboring processors a points at a time The combination of these three effects generated the hundred-fold slow-down The final form of the iPSC algorithms involve no exchange of messages with the host, and buffered all boundary data to avoid sending single points. This should be taken as a very serious demonstration of the importance of coupling algorithm and hardware design We have not had an opportunity to return to Caltech to repeat the old experiments with our optimized software. Undoubtedly efficiencies would be much closer to optimal in all cases with the new software It is also difficult to compare the machines because of the different memory sizes. The largest problems that can fit in the Caltech Hypercube are relatively small iPSC problems, and as a result suffer more from communication startup effects than larger iPSC problems would

1.5. Overview of Contents

In Section 2 we review the hardware and software environment of the Caltech and Intel Hypercubes. In section 3 we introduce two hypercube simulators we have developed that were crucial to the work described in this paper. In section 4 we describe the organization of the hypercube network into various topologies, including grids, trees and a hierarchy of canonical rings Section 5 introduces the library of vector and matrix routines used in most of the further sections and describes a library of useful communication primitives for the Caltech Hypercube Section 6 concentrates on the Vector Library introducing distributed vector formats and the operations defined on them - in particular *inner_product* and *shift_vector*. Section 7 describes the various distributed matrix representations and the Matrix Library, including the important *transpose* and *matrix_vector* operations

The second half of the paper deals with applications of the software described in sections 3-7 In section 8 we introduce the concept of Transpose-Split Algorithms, indicate their implementation on hypercubes and an analysis of their communication vs computation costs. Section 9 discusses the solution of hyperbolic equations using several parallelization schemes Section 10 presents the analysis of computational efficiency as a function of work per grid point and of grid size, referred to earlier.

Section 11 serves as an introduction to the elliptic equation solvers in the remaining sections. Section 12 presents an implementation of a full multigrid elliptic equation solver for hypercubes Section 13 describes a parallel preconditioned conjugate gradient solver for arbitrary-order finite element (or finite-difference) equations Finally section 14 provides an application of Transpose Splitting to the development of a fast solver for the Poisson Equation, which also serves as a parallel preconditioner for the conjugate gradient solver.

In each of the implementation sections we present results of actual hypercube implementations, along with analyses of efficiency and communications overhead

2. ARCHITECTURE AND SOFTWARE

2.1. The Caltech Cosmic Cube

The Cosmic Cube is a parallel processor developed by Geoffrey Fox and Charles Seitz[9,10] at Caltech   The Caltech Mark II Hypercube consists of 2^D ($D = 5$ or 6) independent processors, each with its own local memory There is no shared memory available - the processors cooperate by message passing   Messages are passed over an interconnection network which is a *hypercube* in a space of dimension D, see Figure 1   Processors are located at the vertices of the D-dimensional hypercube and adjacent vertices of the cube are connected by a communication channel along the corresponding edge   All data exchange between processors occurs in 8-byte packets along these cube edges which are asynchronous full duplex channels   In addition to the 2^D node processors, there is a host processor which acts as a control processor for the entire cube and also provides the interface between the cube and a user   All I/O to and from the cube must pass through the host, which is connected to one corner of the cube by an extra communication channel   The original Caltech design consists of a 64-node 6 dimensional hypercube utilizing Intel 8086/8087 processors with 128KB of memory at each node   This architecture has the advantage of being easily fabricated from standard components, and may be scaled up to much larger sizes (in powers of 2) with almost no change in design   Because of these features, machines of this type are likely to become widely available in the immediate future, whereas development of highly parallel global memory machines will take substantially longer

A more advanced Caltech cube called the Mark III is now under development. This will have much faster processors at the nodes (Motorola 68020) and local memory per node will reach several megabytes. Other enhancements will be incorporated based on the experience with the prototype

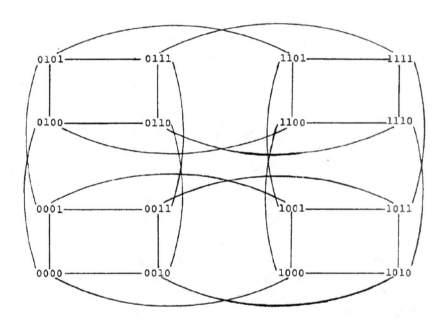

Fig. 1: Representation of a 4 Dimensional Hypercube.

2.1.1. Caltech Hyper-Cube Programming:

There are two fundamentally different communication modes available on the hypercube. In the *Interrupt Driven Mode*, processors are interrupted by messages arriving from the communication channels These messages are preceded by sufficient identification and destination information so that the processor can either forward them to another channel (if the current processor is not the destination) or process the incoming message (if the message has arrived at its destination). In the *Crystalline Operating System* messages are not preceded by address information. As a result, each processor has to know in advance exactly what communication pattern to expect The latter system is unquestionably more efficient, although it is clearly also more restrictive. For the computations described in this paper the Crystalline Operating system was quite adequate The parallelization of other algorithms (e g the local grid refinement algorithms discussed in our related papers[4,6]) will likely require some interrupt driven communication protocols. For the remainder of the discussion we will refer only to the Crystalline Operating System when discussing Caltech Hypercube software.

The software for the cube consists of an operating system kernel, a copy of which resides in each processor, as well as a run-time library providing user access to the communication facilities. Typically, identical copies of a user program are down-loaded to all processors where they execute concurrently. All scheduling is accomplished through communication calls, so that some care is required to prevent locking situations from occurring

As discussed previously, the D-cube has 2^D vertices with D edges radiating from each Thus each processor sees D channels connecting it to its neighbors The cube nodes are numbered in the range $[0, 2^D - 1]$, such that the D-digit binary representations of physically adjacent nodes differ only in 1 bit The channels emanating from a node may then be numbered $0, 1, , D-1$ according to which bit differs in the binary node representations at either end of the channel. There is also an extra channel from node 0 to the intermediate host (referred to as the IH below) through which all communications to and from the cube pass Data to be communicated between processors is sent in 8-byte packets, which are sufficient to encode all scalar data types A set of system calls are available to node-resident programs which implement the required communication primitives for these packets Similar system calls are available on the host to provide communication with the cube In order to give the flavor of these system calls, we list some of the most important ones, along with their functions, in Table 1. In the table the notations *IH* and *ELT* are used respectively for the intermediate host and for the nodes, *data,* represents an 8-byte item, *CUBE* denotes the channel to or from the IH to the hypercube and *P* is the number of processors.

Table 1: Crystalline Cube Communications	
wtIH(data,CUBE)	*Write a data packet from the IH to node 0*
rdsig(data)	*Data sent from the IH is read by each node*
wtres(data)	*Data is sent from a node to the IH*
rdbufIH(datas,CUBE,P)	*Read the union of data sent by all nodes to the IH into array datas*
wtELT(data,chan)	*Send data to the cube neighbor on channel chan*
rdELT(data,chan)	*Read data from the cube neighbor on channel chan*
shift(ind,inc,outd,outc)	*Write buffer outd onto channel outc, then read from channel inc into buffer ind*

The *shift* operator allows either the in or out channel to be specified to be a non-existent channel, denoted NULLCHAN In this case *shift* simply omits the corresponding read or write operation. This feature is especially useful in the treatment of non-periodic grid boundaries where a read or write beyond the boundary is not desired We have found that most hypercube communication may be written in terms of the *shift* operator alone In part this is because the *shift* operator actually includes the *rdELT* and *wtELT* operations which are special cases corresponding to setting *outc* = NULLCHAN and *inc* = NULLCHAN respectively.

One additional routine is very useful in the simulation of many physically interesting problems - such as those derived from discretizations of partial differential equations on regular grids An important feature in such discretizations is that there is typically only nearest-neighbor connectivity among the variables of interest For efficient use of the hypercube, it is then very desirable to map the grid onto the cube in such a way that neighboring grid points (in two or three dimensional space) are mapped onto adjacent nodes of the

cube   Communications overhead will be minimized by such a mapping
Accomplishing such a mapping is non-trivial and in general impossible - for
example there is no such mapping of a 3 dimensional grid onto a 5-cube since
the grid requires a local connectivity of 6 at each node   A general purpose
routine called *whoami*() has been developed by John Salmon at Caltech[11]
based on *binary gray codes,* which generates a suitable mapping of the above
type in most cases where one is possible   The *whoami*() call is usually executed at the start of any grid-oriented program, and in addition to creating a
suitable mapping of the grid to the cube nodes it returns communication
channel information for each of the *grid neighbors* of each processor   This
allows the programmer to think entirely in *grid space* rather than in the less
intuitive *edge space* of the cube

   A hypercube program consists of two separate programs. an *Independent
Host Program* and an *Element Program.* The Element Program, identical
copies of which are executed in all processors of the hypercube simultaneously, consists of distributed data structures such as vectors, matrices, grids
and the parallel operations defined on them. The Independent Host Program
controls the execution of the Element Programs by sending initialization and
control instructions to processor 0, which then broadcasts these to all the
other processors, and by receiving results from the node processors for printing or archiving.

2.2.   The Intel iPSC Computer:

   The Intel Corporation has recently marketed the first commercial realiza-
tion of the hypercube design, based largely on the Caltech Cosmic Cube.
The machine, known as the iPSC, comes in three models - the d5, d6 and d7.
These have respectively 32, 64 and 128 processors   The individual processors are the Intel 80286/80287 with up to 512Kb of memory, and the inter-
connections are provided by high-speed Ethernets, using an Intel Ethernet
chip   The intermediate host machine, which is both the control processor and
the user interface, is an Intel 310 microcomputer running a UNIX system
(Xenix). In addition to the Ethernets along cube edges, a global communication channel is provided from the intermediate host machine to the individual
processors   This feature is useful for debugging and to a limited extent for

control purposes Besides the UNIX system on the host, software for the system consists of a node-resident kernel providing for process creation and debugging along with appropriate communications software for inter-processor exchanges, and for host to processor direct communication Combined computing power of a 128-node system can be over 5 MFLOPS, which along with the 64 Mbytes of memory available, provides a relatively power-ful computer.

2.2.1. iPSC Programming

The software environment for the Intel iPSC is distinctly different from the Crystalline Operating System described above To begin with, the operating system supports multiple processes at each cube node, identified by their process identity number *pid* All communication primitives can address an arbitrary process on an arbitrary node The underlying message passing system includes automatic routing of messages between any two processes This frees the user from developing complex routing schemes in his software, but at the expense of some extra communication overhead We present a list of the iPSC communication calls in Table 2

A further flexibility is the availability of both synchronous and asynchro-nous communication modes The system supports a concept of *virtual chan-nel,* unrelated to the physical channels connecting nearest neighbor nodes A process can communicate with several other processes simultaneously by opening several virtual channels (using the routine *copen*()) and then exchanging messages using asynchronous communication calls All messages have a user-defined integer attribute, called *type,* which is assigned by the sender. A receiver may request messages by type, but not by source process or source node Fortunately the range of the type attribute is large enough ([0,32767]) to allow the source of a message to be encoded in its type. Mes-sages of any size up to 16384 bytes may be sent, although the overhead for message transmission severely discourages sending small messages, a point which we return to in a later section To send a message the message pointer and length are supplied along with the destination node and process, and the *type* attribute (*send*() and *sendw*()) To receive a message, a type and a mes-sage buffer and desired length are supplied, and on receipt of the message the

Table 2: iPSC Communication Routines

Node routines

chan = copen(pid)	open a virtual channel for process *pid*
send(chan,type,mesg,len,node,pid)	send a *type* message to *pid* on *node*
recv(ci,type,msg,len,cnt,node,pid)	read message of type *type*
length = probe(chan,type)	are there messages of type *type* ?
status(chan)	is channel free yet?
flick()	non-busy wait
sendw(chan,type,mesg,len,node,pid)	blocking send to *pid* on *node*
recvw(ci,type,msg,len,cnt,node,pid)	blocking read of *type* *type*
syslog(pid,string)	print a message on the host

Host routines

sendmsg(chan,type,mesg,len,node,pid)	blocking send to *pid* on *node*
recvmsg(ci,type,msg,len,cnt,node,pid)	blocking read

actual length, source node and source process identity *(pid)* are returned
(recv() and *recvw())*. To support asynchronous transmissions, it is possible
to determine if a previous message has completed on a specific virtual channel (with *status())*, and to determine if there is a message of a specific type
pending at a node (with *probe())* One very unfortunate feature of the system is that the host communication primitives are less powerful   For example it is not possible to request receipt of a message at the host by *type* and in
fact one is forced to read messages in totally arbitrary order   As a conse-
quence, it is usually necessary to develop a software buffering scheme for
messages at the host so that incoming messages may be stored until one of an
appropriate type arrives   This is a serious inconvenience and will hopefully
be remedied in future versions of the iPSC software.

2.2.2. Computation and Communication Costs

Two characteristics of the current iPSC design are the slow communication rate and the high overhead for short messages In fact messages of length 0 and 1024 bytes take essentially the same time As a measure of the slowness we note that a message of length 16384 bytes takes 12 seconds to traverse a nearest-neighbor ring of 128 processors, or over 17 seconds using a ring in random (sequential) order The cost of sending a message of length 1 byte to a neighboring processor is approx 5.3 ms while longer messages require about 5.5 ms per 1024 byte segment Figure 2 displays the cost of sending packets as a function of packet size These numbers are approximate and were obtained by sending 30 consecutive messages from node 0 to its 6 neighbors on a 6d cube. An uncertainty is involved because the system clock is accurate only to 16 ms. Larger numbers of messages could not be sent to obtain better statistics without a time delay due to an operating system bug. Messages were sent using both the asynchronous *send* routine and the synchronous *sendw* routine, though there was little difference as indicated by the pair of curves in Figure 2 This slow communication speed is way below the hardware limits of the Ethernet connections and suggests that much time is wasted in operating system overhead Despite this fact we have found that the iPSC can be used with high efficiency on a wide range of problems because of the substantial memory available per node To indicate the processor speed, we note that a C *for* loop with no body requires about 11 micro-secs per point, while a loop with a typical floating point computation such as $a = a + b*c$ requires about 67 micro-secs per point Thus we rate the processor at about .03 Mflops though this estimate might vary by a factor of about 2 in different situations. We summarize processor speed characteristics in Table 3

We present a study of timing costs for messages of different sizes on a 128 processor d7 in Figure 3 The times measured are for passage of messages of various sizes around a complete ring of 128 processors Each processor reads a message from its clockwise neighbor on the ring, then sends the message to its anti-clockwise neighbor. The sequence starts and ends at processor 0 which records the time between sending the message and final receipt The upper curve is for a ring of non-nearest neighbor processors, ordered as 0,1, , 127, while the lower curve is for a nearest-neighbor ring Superimposed on the lower curve is a second curve, generated on the iPSC

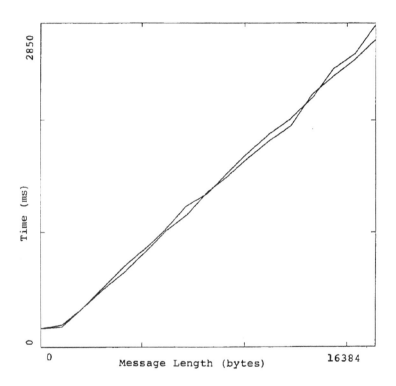

Fig. 2: Synchronous and Asynchronous Message Sending Time as a function of the Message Length on the iPSC.

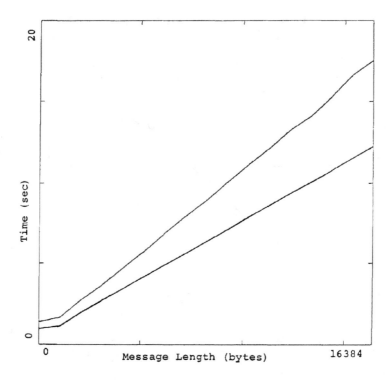

Fig. 3: Time for different size messages to travel around a ring in a 128 node iPSC. The top curve is for the ring in which the processor numbers are increasing. The bottom curve is a nearest neighbor ring (based on Binary Reflected Gray Codes). Superimposed on the lower curve are the times obtained by using the Caltech simulator on the iPSC for the nearest neighbor ring.

Table 3: iPSC Performance	
C for loop· empty body	10 9 μs per point
C loop to copy real numbers	15 7 μs per point
C for loop $a = a + b*c$	67 4 μs per point
send 0 bytes	5.3 ms
send 1024 bytes	5 9 ms
send 2048 bytes	11 2 ms
send 16384 bytes	90 ms
clock() system call	155 μs
clock() time unit	16 ms

using Caltech Hypercube communication calls. The latter calls were implemented using a simulator we have written for the Caltech Hypercube on the iPSC hardware, see section 3 2 As can be seen, overhead for using the simulator is completely negligible

2.3. Basis for Communication Cost Analyses

In subsequent sections, we analyze communication costs for many algorithms Generally we assume that the cost to transfer a segment of k real numbers between two neighboring processors is of the form:

$$ST(k) = \alpha + \beta k \ .$$

This is accurate for the Caltech Hypercube, but is a simplification for the Intel iPSC since the formula does not model the communication cost correctly over the whole range of permissible message lengths. From Figure 3 we notice that messages shorter than 1024 bytes (256 reals) all take essentially the same time This is an important case which we have included in our analyses by using different values α_{long} and β_{long} for long messages, and α_{short} and β_{short} for short messages

We have derived estimates for the coefficients α and β from the data displayed in Figure 3 The data for the lower curve in this figure was obtained by measuring the time necessary to send a message around a 128 node nearest neighbor ring The timing represented there is then the curve $T = 128\ ST(k)$. From this we have deduced that, with times measured in microseconds,

$$\alpha_{short} = 6625\ , \quad \beta_{short} = 8\ 28\ ,$$
$$\alpha_{long} = 3477\ , \quad \beta_{long} = 22.5\ .$$

These numbers are in sharp contrast with the cost γ to perform a typical arithmetic operation, which from Table 3 is seen to be of order 30 microseconds. In particular the ratio $\alpha_{short}/\gamma = 220$ indicates that communication of single data items is hundreds of times slower than a corresponding computation. Another parameter that appears in the analysis of some algorithms is the length λ words of a buffer used to accumulate short messages for communication in a single packet Ideally λ should be chosen such that $\alpha/\lambda < \beta$. On the iPSC we have used $\lambda = 4096$

For comparison we present here the corresponding data for the Caltech Hypercube In that case a single value of α and β suffice to cover the whole range, and we find·

$$\alpha = 92\ , \quad \beta = 40\ ,$$

Thus message startup overhead becomes small as soon as even a few words are communicated between nodes, although communication rates for very long messages are about twice as slow as on the iPSC Computation rates γ for the Caltech Hypercube are comparable to those for the iPSC The buffer of length λ is not required at all.

The four parameters α, β, γ and λ appear in efficiency estimates throughout the paper without further comment We also use γ as a measure of integer computation speed, for example it is used to measure the cost of copying arrays In addition we use N to denote the size of vectors or the dimension of arrays and we use $P = 2^D$ to denote the number of processors.

3. HYPERCUBE SIMULATORS

An important aspect of our hypercube work was the development of a hypercube simulator for the Intel iPSC. It is difficult to develop programs for a parallel computer such as a hypercube in a reasonable period of time without access to a simulator for the machine. A simulator allows programs to be developed on faster and more robust computers, tested under conditions of moderate parallelism (we have used 4 to 16 processor simulations) and then moved to the target hardware with reasonable assurance that issues of synchronization and control are correct. One can then concentrate on the many remaining issues that can prevent successful execution, such as faulty node compilers, defective processor hardware or bad communication lines. In addition, by instrumenting a simulator one can obtain useful statistics when real codes are run. We have instrumented all of our codes, and statistics on a per-processor basis are collected on many items such as numbers and sizes of messages sent and received, amount of memory used and so on. This information is extremely helpful in pin-pointing communication bottlenecks in the code. These statistics are collected by each node separately and are automatically sent on program termination to the host machine for storing in a history file. We have also instrumented the standard communication calls used on the real hypercube so that history files are obtained during actual runs as well.

We describe now two hypercube simulators we have developed, one for the Intel iPSC, and one for the Caltech Hypercube. These simulators may be run across a series of machines, with sets of nodes assigned to different machines. This allows for much faster simulations. We have also simulated the Caltech Hypercube on the Intel iPSC, which allows Caltech Hypercube programs to be ported directly to the iPSC without any code modification.

3.1. An Intel iPSC Simulator for UNIX:

We have developed an Intel iPSC simulator that uses the UNIX operating system to simulate both hypercube processes and inter-process communication. The simulator actually represents a general model of message-based parallel computation and can be used to simulate various different architectures We use the operating system to implement processes, the directory tree to implement nodes and the file system to implement message passing Each hypercube node is assigned a sub-directory of the running directory, with the directory name being the hypercube node number. The hypercube host is assigned to a host subdirectory Each process running on that node is assigned to a sub-subdirectory, again numbered in relation to the number of processes on the node Thus process *pid* on node *node* resides in directory */node/pid*.

The simulator starts by creating all of the relevant directories and starting up the requested programs in the appropriate directories. The processes then communicate with each other, and with the host, by exchanging files containing their messages The procedure here is simple· a process creates a message in a message file with a name *source seq_num* that identifies its source and also a *sequence number* from the source to the destination The sequence number is incremented each time the source sends a message to that destination Each process records the sequence number of the last message sent to each of the other processes, initialized to 0 Furthermore each process remembers the next expected message sequence number from each of the other processes, again initialized to 0 Once a message is complete it is moved to the appropriate destination directory The receiving node looks for appropriately named files in its directory and on reading a file, adds the message to a linked list of messages. This list is ordered by source node, and within that by sequence number. When a message is requested of a particular type (by a *recv* or *recvw*), the list is searched for a message of that type whose sequence number is the next expected one from its source. This ensures that messages from the same source are received by programs in the order sent. If no such message is found, the process continues reading and storing incoming messages until a suitable one appears

The simulator is independent of the mechanism for message routing Thus various message routing systems may be added to simulate actual hardware designs The simplest message routing uses file renaming to transmit a file to its destination directory. This simulates a fully-connected network hardware in that all processes are connected to each other in a completely equivalent fashion We have also used the *socket*()-based interprocess communication facilities of 4.2bsd UNIX to implement message transmission as an alternative to file passing. However the file mechanism has the advantage of working over networks of heterogeneous systems and with various flavors of UNIX. In addition to the fully-connected network, we have also simulated a hypercube network. In this case messages are passed through a sequence of *nearest-neighbor directories* in order to reach a target. We stress that for most purposes this is completely unnecessary and significantly slows the simulation since most messages must be handled several times. In fact the fully-interconnected network is perfectly adequate for iPSC simulations since the iPSC software supports symmetrical communication among all processes, completely hiding the hardware.

3.2. A Caltech Hypercube Simulator on the Intel iPSC:

Developing a Caltech Hypercube simulator for the iPSC also proved to be useful. This allows identical codes (ie Caltech Hypercube codes) to be run on both machines, with a software overhead that we have measured to be under 1%

Caltech processes are implemented as iPSC processes, with only one per node allowed. Since Caltech processes work with channel numbers, we first create a mapping of channel numbers to node numbers. The basic communication primitives such as *rdELT*() and *wrtELT*() (see Table 1) are then implemented using the iPSC *recvw*() and *send*() primitives (see Table 2). Note that Caltech protocols of reading and writing channels imply that when receiving a message the message is requested from a specific source, the node at the other end of the requested channel. On the other hand iPSC protocols do not allow for requesting receipt of messages by source location, but only by message type. We overcome this problem by using the iPSC message type to encode the source node of a message when sent. A receiver then requests a message with type equal to the desired source node number. The only complexity encountered is at the host processor. At the host, the iPSC protocol does not even allow asking for messages of a specific type - one simply reads messages as received. This issue is dealt with using a complicated buffering scheme where incoming messages are stored and ordered with respect to both sequence number and type. A Caltech Hypercube simulator for UNIX was previously written at Caltech[12] but the one developed here is different in that it is targeted at another parallel machine - the iPSC By combining the Caltech Hypercube simulator for the iPSC with the iPSC simulator for UNIX, we have created a new Caltech Hypercube simulator for UNIX

4. HYPERCUBE INTERCONNECTION TOPOLOGIES

The interconnection network of the hypercube can be organized in a variety of ways depending on the application   In the following subsections we consider some of these topologies since they are essential for many of the later algorithms. The cases of most importance to us are mappings of grids, trees and rings to the cube network

4.1. Grids on Hypercubes

Grids are mapped onto the Hypercube using the *whoami* facility developed by J Salmon [11] The routine uses Binary Gray Code (BGC) sequences[13,14] to number processors. These are sequences of numbers such that consecutive terms in the sequence differ by only one bit in their binary expansions. For example, one 4-bit BGC is the sequence (by rows)·

0000 0001 0011 0111 1111 1110 1100 1000
1001 1011 1010 0010 0110 0100 0101 1101

As mentioned earlier, nodes on the hypercube with processor numbers that differ in only one bit in their binary expansion are physical neighbors. If we interpret the terms in a BGC sequence as processor numbers, then nearest neighbors in the sequence will also be physical neighbors   This allows one dimensional grids to be mapped onto the hypercube using only nearest-neighbor connections. The generalization to higher dimensional grids is done by mapping each dimension to a suitable subcube and using the 1-dimensional mappings for those sub-cubes.

Certain applications require periodic grids to be mapped to the cube   The above BGC sequence is not suitable for this purpose because its last element is not a neighbor of the first one. A sub-class of Gray codes, the Binary Reflected Gray Codes (BRGC), have the desired property [11] In the sequel we will always use a canonical BRGC defined recursively as follows With one bit the BRGC sequence is   0 1   With d-bits we take the $d-1$ bit sequence, followed by the same sequence reversed in order and with each dth bit set. Thus with two bits we have the sequence 00 01 11 10 · and with four bits the sequence is:

$$0000\ 0001\ 0011\ 0010\ 0110\ 0111\ 0101\ 0100$$
$$1100\ 1101\ 1111\ 1110\ 1010\ 1011\ 1001\ 1000$$

This is the standard processor ordering used by the Caltech Hypercube research group [12] With a BRGC ordering the subcubes used for a grid whose dimensions are each a power of 2 will all be periodic, allowing periodic grids of such dimensions to be mapped to the cube. The BRGC sequence also has the property that two elements that are a power of 2 apart differ in at most two bits [15]

4.2. Trees on Hypercubes:

There are several ways to embed a tree onto a hypercube. The most natural trees are either binary trees or D-ary trees where D is the cube dimension. In the following discussion we choose the root of the tree to be at node 0, but any other point can be used as easily.

The observation that allows a balanced binary tree to be mapped to the hypercube, is that a D-cube may be regarded as two $(D-1)$-cubes with corresponding processors from each connected by an extra channel. Numbering the processors in the two $(D-1)$-cubes as p_0, ., $p_0+2^{D-1}-1$, and p_0+2^{D-1}, ., p_0+2^D-1, respectively, we connect processor p_0+p in the first subcube with processor p_0+p+2^{D-1} in the second subcube. The binary tree on the D-cube is then defined recursively as having the lowest numbered processor as its root and the two $(D-1)$-cubes as its left and right sub-trees Note that this is a logical binary tree only - some processors occur several times at tree nodes. For example, processor 0 occurs D times, being the root of D sub-trees Each processor is located exactly once at a leaf of the tree.

An alternative tree mapping is to represent the network as an unbalanced D-ary tree. This mapping is based on the obvious fact that D connections emanate from every node This tree is a physical tree - each node in the tree corresponds to a unique processor The tree may be defined either by giving the children of each node, or by specifying the parent of each node. The children of a node are the processors whose node numbers are obtained by setting in turn each of the low-order unset bits in the D-digit binary representation of the node number. Alternatively, the parent of a node is located by unsetting the lowest-order set bit in the binary representation of the node number

These two logical trees are in reality two different ways of representing the same set of connections. Depending on the application it can be helpful to consider trees from either viewpoint Figure 4 represents the 4-cube as a tree from these two points of view

4.3. The Hypercube Network as a Hierarchy of Rings.

As indicated above, it is possible to map a ring structure onto the hypercube by giving the processors the position of their processor number in a Binary Reflected Gray Code sequence As a property of the binary gray codes, processors that are neighbors in this ordering (logical neighbors) are also neighbors on the Hypercube (physical neighbors). We will use the terms *logical distance* and *physical distance* in this sense throughout the rest of the paper. As noted in section 5.1, the BRGC ordering has the further property that two processors at logical distance 2^d, $d = 1$, ., $D-1$, in the ring are at a physical distance 2 For each d in the range $1, \ldots, D-1$ there is then a collection of subrings of processors such that processors at logical distance 2^d in the subring are at physical distance 2 We call these the *level d subrings* of the hypercube. Every processor is on exactly one subring of level d A subring connecting processors at logical distances 2^d in the main ring contains 2^{D-d+1} nodes, and hence there are $2^D / 2^{D-d+1} = 2^{d-1}$ such subrings covering the hypercube. In Figure 5, we display on the left the 4 logical hierarchies of rings in a 16 node (4 dimensional) hypercube which has been ordered with the BRGC. On the right, we show the actual physical communication channels used to implement the logical structures. Note that except for the level 0 ring, logical neighbors on the subrings always are separated by distance 2. We see that the main ring structure is used for the rings at levels 0 and 1. Level 2 consists of 4 logical subrings The physical links at level 2 however form only 2 subrings, corresponding to the fact that 1 logical link requires 2 physical links. It should be noted that parallel communication is possible on all subrings at the same level.

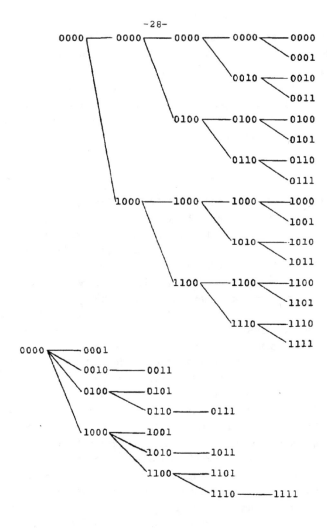

Fig. 4: Nodes of a 4 Dimensional Hypercube as a Balanced Binary Tree and as an Unbalanced Tree of order 4.

LOGICAL STRUCTURE   **PHYSICAL STRUCTURE**

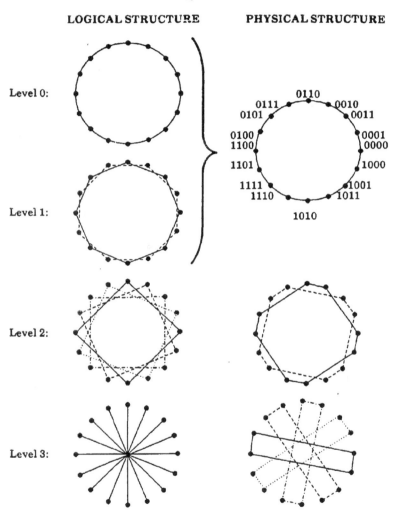

Fig. 5: The 4 logical hierarchies of rings in a 16 node hypercube and their communication channels used to implement them.

We have implemented a library subroutine that, analogously to the *whoami* routine mentioned earlier, returns the channel information needed to use the hypercube as a hierarchy of rings. This routine starts by creating a level 0 ring, by calling the *whoami* routine to construct a one-dimensional periodic grid. From this ring the other levels can then easily be constructed On the Caltech Hypercube, the information returned consists of $D-1$ pairs of channels: a left and a right channel for each physical ring level. This gives a complete description of the subring hierarchy since each processor is a member of just one physical subring at each level

5. PORTABLE PARALLEL LIBRARIES

In this and the following two sections of the paper we introduce and describe the various software libraries that are the core of our parallel implementations for hypercubes. These libraries will be used in the later application sections for solving partial differential equations. Section 5 1 discusses a simple object-oriented communications library which greatly reduces the tedium of packaging quantities into 8-byte arrays on the Caltech Hypercube. Section 6 introduces the Parallel Vector Library, with subsections discussing specifics for the more complex vector operations such as inner product and shift-vector, while section 7 introduces the Parallel Matrix Library, again with subsections treating the matrix-vector and transpose operations

5.1. An Object-level Communication Library

The Caltech Hypercube communication software deals with messages in terms of 8-byte units. This tends to lead to very ugly code where 4 integers, or 2 floating point numbers or the like are stored into an 8-byte array for transmission Resulting code is hard to read and prone to bugs, especially if the receiver decodes the bytes differently from the sender's encoding. On the Intel iPSC we have seen that sending lots of short messages is extremely inefficient.

We have developed a simple higher-level interface to these routines that allows objects (data-structures) of arbitrary size to be communicated between nodes and with the host with essentially optimal efficiency In Table 4 we indicate the calling sequence of some of these routines. In addition to sending linear arrays and structured types (as in C) between processors, this library also supports communication of vectors and matrices between processes and with the host In particular there are routines to send a globally distributed vector or matrix to the host, frequently used in printing the final results of a program. The *send_local_vectors()* routine allows r vectors $v_1, \quad ,v_r$, with addresses stored in the array v and with lengths $n_1, \cdots ,n_r$ stored in the array n to be sent to a neighbor. On the Caltech Hypercube this routine simply sends each vector in short segments, whereas on the Intel

---

Table 4:   Communication Routines.

send_structure_to_cube(struct size)

read_structures_from_cube(structs,size)

send_structure_to_host(struct,size)

send_structure_to_node(chan,struct,size)

recv_structure_from_host(struct,size)

recv_structure_from_node(chan,struct,size)

send_local_vectors_to_node(chan,r,n,v)

recv_local_vectors_from_node(chan,r,n,v)

send_matrix_to_cube(mat)

recv_matrix_from_host(mat)

send_matrix_to_node(chan,mat)

recv_matrix_from_node(chan,mat)

---

iPSC the vectors are first copied and concatenated into an array of length 4096 bytes which is sent each time it fills with a single *send* call The last segment sent may of course be shorter Unless all of the vectors are already over 1024 bytes, this results in much more efficient iPSC communication, at the cost of copying the vectors If some of the vectors supplied to the routine are already long, then they are sent directly to avoid the extra copying overhead. The *recv_local_vectors*() routine just splits and copies incoming 4096-byte segments into r local vectors. Similarly the *send_matrix_to_node*() routine buffers the rows of the matrix into large messages on the iPSC

The only non-trivial issue in the other routines is the handling of quantities whose size is not a multiple of 8 bytes In these cases it is necessary to send a last packet containing some zero data in order that the full communication be a multiple of 8 bytes On receiving messages this can require buffering the message first, and then copying it to its destination - in particular this occurs on reading the union of messages sent by all nodes to the host, if the messages are not a multiple of 8 bytes For example, the *read_structures_from_cube* routine reads a structure from each processor into a contiguously stored array of structures supplied by the caller   The array cannot be used to read the incoming structure messages since these are in fact larger (if not multiples of 8 bytes) than the structures in the array

The matrix operations in the communication library depend on the format for representing distributed matrices, which we will describe in detail in section 7.

6. PARALLEL VECTOR OPERATIONS

We describe the development of a library of vector operations on a hypercube. This description is valid for a wide range of parallel processors, including shared memory and even serial machines. All that we assume is that the processors may be organized into a one-dimensional *ring* of processors in which each processor can communicate efficiently with its neighbors on either side On a hypercube there are many ways to imbed such a ring onto the processor network in such a way that only nearest neighbor connections are involved - see section 5 In order to simplify notation, we assume in the following that the processor ordering 0, 1, $P-1$, 0 is already such an optimal ring

An important feature of this library is that individual processors work entirely with a local vector and are never aware of the location of other parts of the distributed vector This is true even with operations such as *inner_product*() which involve communication between different processors Furthermore the operations on local vectors are themselves vector operations Thus the library could take full advantage of an available vector processor on each node

6.1. Distributed Vectors

We use a consistent scheme for distributing all vectors over the P processors. We consider here only vectors of length $N \geq P$. We divide the vector into P contiguous and roughly equal-length segments It is then possible to distribute a segment to each of the P processors We will call a vector stored in this way, a *distributed vector*. We will assume that every distributed vector of the same length is distributed in the same way over the processors, i e segments of different vectors residing in the same processor have the same length. The actual distribution formula we have used is as follows. We divide the vector length N by the number of processors·

$$N = hP + r, \quad 0 \leq r < P$$

The first r processors will be assigned segments of length $h+1$ elements and the remaining $P-r$ processors will be assigned segments containing h elements   In the case that P divides N exactly, all segments are then of size $h = N/P$

To create a distributed vector of length N, all processors call an allocation routine·

$$v = allocate\_vector(N, type)$$

Several types of vector representation are supported, of which the most important types are called *SIMPLE* and *SHIFT*   The *SIMPLE* form represents each vector segment as an array of length h or $h+1$. In addition to performing the above computation for the segment length in each processor, this routine allocates storage for the array and returns a pointer to a data structure v which represents the vector. The data structure records the type, local segment length and a pointer to the start of the vector   In the *SHIFT* format, each vector segment is located in a buffer of length $2h$ within which it is allowed to slide around (see the following section on shifting vectors)   Initially the segment is centered in the buffer. In this case, the data structure v records the type and local segment length, pointers to the start of the segment both before and after shifting, as well as the buffer extremities. Whenever it is intended to apply the *shift_vector* operation to a vector, the vector should be allocated of type *SHIFT*. A library routine *convert_vector()* is available to convert a vector from one type to another

6.2. Operations on Distributed Vectors.

We have implemented a large set of operations on such distributed vectors. Application of any such operation is accomplished by having all of the processors call the appropriate routine simultaneously, with the appropriate vector data structures as arguments   Vectors of different types may be freely mixed, as appropriate conversions are applied when needed   We list sample routines from the library, along with their purpose, in Table 5

Most of these routines are trivial to parallelize. However· *shift_vector*, *inner_product*, *max_vector* and *sum_vector* involve communication between processors. We discuss these four routines in more detail in the following subsections   The remaining routines involve no communication are are

Table 5. Basic Vector Routines

allocate_vector	*allocate local segment of a vector*
delete_vector	*deallocate storage* for *a vector*
convert_vector	*change type of vector*
zero_vector	$v_i = 0$
shift_vector	$v_i = v_{(i-s)mod V}$
max_vector	$\displaystyle\max_{0 \le i < V} v_i$
sum_vector	$\displaystyle\sum_{i=0}^{V-1} v_i$
inner_product	$\displaystyle\sum_{i=0}^{V-1} u_i v_i$
copy_vector_to_vector	$v_i = u_i$
add_scalar_times_vector_to_vector	$u_i = u_i + sv_i$
add_vector_to_scalar_times_vector	$u_i = su_i + v_i$
add_vector_times_vector_to_vector	$u_i = u_i + v_i w_i$

optimally parallel· the corresponding operation is performed in each processor on the local segments of the vector arguments. using the standard serial code. Thus we will not discuss routines such as *add_vector_times_vector_to_vector* any further

6.3. The Vector Shift Algorithm

The *shift-vector* operation consists of sliding each element of the distributed vector forward by s elements, placing the last s elements of the vector at the start. It is essential that this operation be very efficient since it is used by many of our other algorithms. For this operation we require that each vector segment is of at least length one, or, in other words, that each processor contains at least one element of the vector

We discuss first the simple case of a short shift of length 1 Most of the shifting will occur internal to a processor If there are h elements in a segment, the cost of sliding the segment forward is $O(h)$ In addition, each segment has to execute communication calls to send its last segment component to the following processor and to receive the last component from the previous processor. All of the internal shifting within nodes may be done in parallel, and similarly for the shifting of the segment ends, so that overall the cost of the operation appears to be $O(h)$, which is unacceptable.

We have implemented the shift operation as an $O(1)$ operation at the cost of using some extra memory in each node The idea is to imbed the local segment of each vector in a larger buffer which extends a distance $O(h)$ (for example $h/2$) on either side of the segment For this purpose, vectors are allocated by the *allocate_vector* routine using type *SHIFT*, see the previous section. Vectors are then shifted by simply leaving them in place, adding the new end element from a neighboring processor to the appropriate end of the local vector. The effect of this is to cause the vector to slide around in its buffer with successive shifts The pointer to the start of the vector in the corresponding vector data structure is incremented for each shift After $O(h)$ elementary shifts, the vector may have slid to the end of its buffer in each node At this point an $O(h)$ operation is required to copy it back to its initial point in each processor. However one can regard this $O(h)$ operation as amortized over the $O(h)$ shifts that preceded it, so that the overall cost of a shift is still effectively $O(1)$. This copy back to the original position is automatically detected and performed by the software When shifted vectors are involved in other vector operations, the appropriate offsets are automatically used for each vector

The shift operation for large shifts, consisting of many one element shifts in succession, can be made even more efficient by using the full hypercube hierarchy of rings at many levels in addition to the top level ring used above As this optimization is somewhat complex we assign the discussion to a separate section.

6.4. The Optimized Vector Shift Algorithm

Based on previous observations, we can optimize the *shift* operation for the case in which the number of shifts is large. In fact we will show that large shifts may be performed $P/\log P$ times faster than if the simple shift algorithm above were used. In the notation of the previous section, we require that $r=0$ for this algorithm, so that the length N of the vector is an exact multiple of the number of processors: $N=hP$. Thus every processor contains a segment of length h of the vector.

We introduce some useful terminology. An *element shift* will denote the logical operation that shifts a vector by one element. An *element transfer* will denote the physical communication operation between two processors required to transfer a vector element from one processor to another. Analogously, we call a *segment shift* the logical operation of shifting a vector h times. A *segment transfer* is then the communication operation between processors to transfer a segment. Note that a segment shift results in each processor transmitting its complete segment to the following processor in the ring, receiving in turn the complete segment of the preceding processor. We assume that an element transfer takes time $O(1)$ while a segment transfer takes time $O(h) = O(N/P)$. We will show that any shift can be performed in time at most $O(N\log P/P)$.

Suppose we need to shift a vector by m element shifts, where without loss of generality we assume $m>0$. Writing

$$m = m_s h + m_e, \quad 0 \le m_e < h,$$

we can reduce the m element shifts to m_s segment shifts and m_e element shifts. Note that if $m_s \ge P$, we can reduce it further by letting $m_s = m_s \bmod P$, since a shift of P segments is a shift of $N=hP$ elements and thus is equivalent to not shifting at all. A further immediate savings is possible by observing that when $m_s > P/2$, it is better to shift the segments in the opposite direction $m_s - P$ times. Suppose that all these reductions are done on m_s. Then the resulting m_s will be in the range $-P/2+1$, . ., $P/2$.

The m_e element shifts are performed separately as in the previous section. In fact they may be lumped together into one for the internal shift, which then becomes an $O(1)$ operation using the buffering mechanism described earlier. Because of the need, mentioned earlier, to occasionally copy a shifted segment back to its initial position, a better estimate of the cost of m_e internal shifts would be that it is $O(m_e)$. The m_e last elements of each segment will be sent to the following segment, resulting in a communication cost that is $O(m_e)$. Therefore the total cost of m_e element shifts is $O(m_e)$.

We now use the hierarchy of rings to speed the segment shifts. Consider the binary representation $b_{D-2} \cdots b_1 b_0$ of m_s. For each $i > 0$ such that b_i is set, we need to perform 2^i segment shifts, which reduce to just one segment shift on the logical sub-rings of level i. As discussed in section 4.3 this may be accomplished with just 2 shifts on the physical subrings of level i. This has the effect of performing 2^i segment shifts while taking the time of only 2 segment transfers. Bit 0 is an exception because if set it requires only one segment shift of the main ring, taking the time of one segment transfer. Thus even in the worst case, where all $D-1$ bits in m_s are set and where the number of element shifts is $h/2$, the total time is at most.

$$T_{shift} = \{2(D-2)+1\} T_{segment} + h/2 \, T_{element}$$
$$= O(\log P) O(N/P) + O(N/P).$$

Note that these estimates neglect any overhead related to sending short messages. In section 7.6.1 we present a more complete analysis of these costs. We summarize these results as two theorems:

Theorem 1: The time $T_{shift}(m)$ to shift a vector of length N by a distance m on a hypercube with P processors satisfies.

$$T_{shift}(m) \leq \alpha(1+d(m)) + 2\beta d(m)N/P + (\beta+\gamma) \, m \bmod(N/P),$$

where $d(m)$ is the number of set bits in the binary expansion of m div N/P.

Theorem 2: Upper bounds for the maximum time T_{shift}^{max} and average time T_{shift}^{ave} to shift a vector of length N by a distance m on a hypercube with P processors are

$$T_{shift}^{max} \leq (2\beta \log P + \gamma)N/P + 2\alpha \log P,$$
$$T_{shift}^{ave} \leq (\beta \log P + \gamma)N/P + \alpha \log P$$

The algorithm above fails if the length of the vector is not a multiple of the number of processors In this case there are r segments of length $h+1$ and $P-r$ segments of length h. With segment shifts, the segments of length $h+1$ are transported to arbitrary processors To bring the vector back to its original form (with the longer segments in the logically early processors) requires $O(N)$ element transfer operations, destroying any efficiency obtained by using segment shifts One solution in such cases is to solve a slightly larger problem such that the length of the vectors is a multiple of P

6.4.1. Timings for the Optimized Vector Shift

We have measured the effectiveness of the optimized vector shift by timing shifts of various sizes for vectors of varying length and with different numbers of processors Figure 6 shows the time required to shift a vector of length 320 on a 32-node iPSC. The graph shows the shift time as a function of the shift length, which varies from 1 to 319. Figure 7 presents the same statistics for a vector of length 1280 on a 128 node iPSC, again shifted by distances 1 through 1279. These two cases are useful to compare since the number of elements h per node segment is the same in each case, namely 10 We note that the cost of performing shifts on the larger cube is only slightly more than on the 32-node cube Specifically the slowest shift in the 128 node case is only about 10% slower than in the 32 node case. Furthermore the variation from longest to shortest shift is comparable Finally in Figure 8 we present shift statistics for a vector 100 times longer - i.e of length 128000 so that there are 1000 elements per node The shift times for the shorter vectors are dominated by the high message start-up cost of the iPSC

6.5. Inner Product

The semantics of the inner product operation in our library needs careful discussion Each processor calls the *inner_product*() routine simultaneously, providing as arguments the two vector segments local to the processor The *inner_product*() routine returns in *every* processor with the value of the inner product of the two complete distributed vectors No processor is

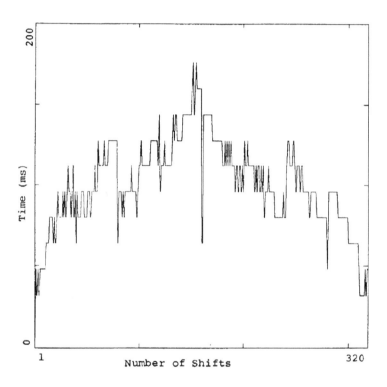

*Fig. 6: Time to shift a vector of length 320, distributed over the
32 node iPSC (10 elements per processor), as a function of the
number element shifts.*

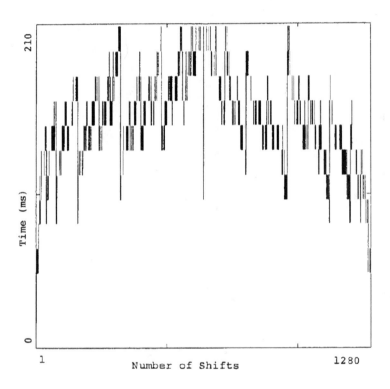

Fig. 7: Time to shift a vector of length 1280, distributed over the 128 node iPSC (10 elements per processor), as a function of the number element shifts.

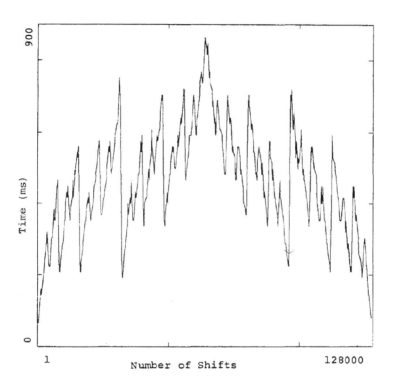

Fig. 8: Time to shift a vector of length 128000, distributed over the 128 node iPSC (1000 elements per processor), as a function of the number element shifts. (Only representative sample is displayed.)

distinguished in these semantics.

The implementation of the routine does of course distinguish processors. The inner product operation is complicated because it involves communication between the processors to sum inner products of the segments local to each processor, and further communication to broadcast the final sum back to all of the nodes. On cube architectures, this should be done by summing over a *tree* of processors, since if one processor is used to accumulate all partial inner products, it will result in a critical section. A D-ary tree is mapped onto the processor network with node 0 as the root. The parent of a node is defined by zeroing the lowest non-zero bit of the node number while the children of a node are obtained by setting in turn exactly one of the low-order zero bits of the node number. The inner product routine then takes the form.

```
real inner_product(v1,v2)

mysum = local_inner_product(v1,v2)

for each child
        read child_sum from child
        mysum = mysum + child_sum
end for

if (node≠0) send mysum to parent
if (node≠0) read sum from parent

for each child
        send sum to child
end for

return sum
```

With this implementation the cost of an inner product will be $O(N/P)+O(\log P)$, with the $O(N/P)$ reduced further if vector hardware is available on the nodes. Clearly all of the local sums may be computed in parallel in time $O(N/P)$, proportional to the local segment length $h=N/P$. The sum of the partial sums over the D-ary tree may be computed in time D, if the unit of time consists of 1 real addition plus communication of a real number to a neighbor. To see this, note that the unbalanced D-ary tree rooted at node 000. 0 contains $(D-1)$-ary , $(D-2)$-ary, ... 0-ary sub-trees rooted at its children 100 .0, 010. 0, 000 .1 respectively. Assume as an induction principle that the result is true for each sub-tree of fanout $0, 1, .. , (D-1)$. The hypothesis is clearly true for $D=0$ and $D=1$ By the induction hypothesis, the sum from node 100 0 will arrive at 000..0 at time $D-1$, from node 010. 0 will arrive at time $D-2$ and so on Each partial sum can then be added to the total in 000 0 while the next partial sum arrives It follows that all summation will be completed by time D, completing the induction. The broadcast of the result to the individual processors reverses the above steps and also requires time $O(D)$.

Following the steps in the proof in detail and using the formulation in section 2.3 for the cost of sending messages, we arrive at the result·

Theorem 3: The cost of performing an inner product of two vectors of length N on a hypercube with P processors satisfies

$$T_{inner\_product} \leq 2\gamma N/P + (2\alpha_{short} + 2\beta_{short} + \gamma)\log P$$

We assign the same semantics to other related operations such as $max\_vector()$ and $sum\_vector()$ Again $max\_vector()$ will be called by each node with the local segment of the vector as argument and returns with the maximum component of the corresponding *global* vector in time $O(\log P)$

In Figure 9 we display the execution time of the *inner_product()* routine for vectors of increasing size, from 1 to 950 elements per processor, on the 128 processor iPSC. The initial point on the curve is related to the high startup cost for short messages on the iPSC Since the communication cost is independent of vector size, the curve is essentially linear reflecting the time taken to compute the local inner products within each node. As can be seen, even for vectors of length over 120000 elements the inner product is dominated on the iPSC by the communication cost

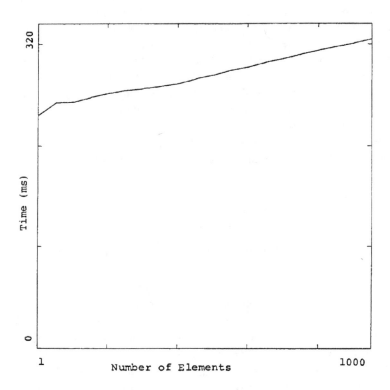

Fig. 9: Time to compute the Inner Product of a Distributed Vector on the 128 node iPSC as a function of the Number of Elements per processor.

7. PARALLEL MATRIX OPERATIONS

Distributed matrices are of special importance in numerical algorithms and we use several different representations for them depending on the application By far the most important routines in this library are the *transpose*() and *matrix_vector*() operations The library also contains routines to allocate and print matrices, to exchange submatrices among nodes, and to convert matrices from one representation to another. We summarize in Table 6 the basic matrix operations of importance. Higher level operations such as addition and multiplication of matrices are easily constructed from the basic set

Table 6: Basic Matrix Routines:	
allocate_matrix	*allocates global matrix in each node*
delete_matrix	*deletes an allocated vector*
convert_matrix	*convert between matrix formats*
send_matrix_to_cube	*host sends a matrix to the nodes*
recv_matrix_from_host	*host reads a global matrix*
send_matrix_to_node	*send piece of matrix to another node*
recv_matrix_from_node	*receive piece of matrix from another node*
transpose	*transpose a matrix in place*
matrix_vector	*global matrix times distributed vector*
areal_parameters	*set areal decomposition parameters*
shuffle	*exchange areal boundaries with neighbors*

As with distributed vectors, distributed matrices are represented by small data structures in each processor that encode the location of the data, the type of the matrix representation, the array dimensions and other relevant information This allows matrices of different types to be handled consistently. Since each processor is aware of the data representation it is working with, it is not necessary to depend on some control processor to handle flow control for matrix operations. The most important representation types are *CONTIGUOUS_ROW*, *DISTRIBUTED_ROW*, *CONTIGUOUS_COLUMN*,

DISTRIBUTED_COLUMN, *DIAGONAL* and *AREAL* The allocation routine has the form.

$$mat = allocate\_matrix(rows,cols,type,nsparse,sparse).$$

If *nsparse* is non-zero it indicates that the matrix is sparse, and the value of nsparse then indicates the number of non-zero rows, columns or extended diagonals according to the value of *type* The argument *sparse* is a pointer to an array of *nsparse* integers indicating which rows, columns or extended diagonals are non-zero. Sufficient storage is set aside to store the matrix, and the data structure *mat* when returned includes the sparsity information as well as the size of and pointer to the local matrix segment stored in each processor The sparse format supported here does not include general sparse matrices, but is adequate for a large class of matrices whose sparsity structure has a diagonal form For type *AREAL* matrices the sparsity information is ignored, but certain other information is obtained in this case from the *areal_parameters*() routine which must be called before any matrix of this type is allocated, see section 7.3 below

7.1. Row and Column Matrix Formats:

The simplest representations of an $N \times N$ matrix are row or column oriented. We make use of two different row representations As usual we decompose N as· $N = hP + r, 0 \le r < h$. In the *CONTIGUOUS_ROW* representation, each row is stored entirely in one processor The first r processors contain $h + 1$ consecutive rows of the matrix, which we store contiguously as an $(h+1) \times N$ matrix, while the remaining $P - r$ processors contain h rows of the matrix, stored as an $h \times N$ matrix In the *DISTRIBUTED_ROW* representation, each row is stored as a distributed vector, with the vector segments in each processor from consecutive rows stored contiguously This representation is fully defined by the distributed vector format discussed previously and we say no more about it here as a result. We note that the N vector segments stored in one processor form an $N \times (h+1)$ or $N \times h$ matrix. The corresponding column representations are obvious

The two representations are in fact closely related Given a *CONTIGUOUS_ROW* representation we can convert it to the

DISTRIBUTED_COLUMN form by simply transposing each of the submatrices stored in individual processors. Similar transpositions convert a *DISTRIBUTED_ROW* format to *CONTIGUOUS_COLUMN* format. Since no communications are involved, and all local transpositions may occur in parallel, the time to effect these transformations is $O(N^2/P)$

7.2. The Diagonal Form of a Matrix:

The third representation of a matrix we call the *DIAGONAL* form. This form is especially convenient for matrices which have a small band-width or which contain only a few non-zero diagonals. We consider an $N \times N$ matrix A with rows indexed by i, columns by j, both i and j in the range $0, \ldots, N-1$. The kth diagonal of A where k is in the range $-(N-1), \ldots, (N-1)$, consists of those elements $A_{i,j}$ such that $j - i = k$. This diagonal contains $N - |k|$ elements. We construct *extended diagonals* D_m indexed by m in the range $0, \ldots, (N-1)$ as follows. For each $k>0$, consider the diagonals k and $k-N$ which have respectively $N-k$ and k elements. It is therefore possible to construct a vector of length N, which we will index by $m=k$ and call an extended diagonal, consisting of diagonal $k-N$ in the first k components of the vector and diagonal k in the last $N-k$ components. The complete matrix A is uniquely represented by the N extended diagonals of length N. For the rest of the paper we will refer to these extended diagonals simply as diagonals.

It is convenient to introduce a matrix D whose rows are the extended diagonals of A (ordered by the index m introduced above). More compactly, we can describe D as an *index transformation* of the matrix A:

$$A_{i,j} = D_{(j-i)\bmod N, j}$$

This notation will be used frequently in the sequel without further comment. Figure 10 displays a 6×6 matrix A and its diagonal form D.

The diagonal form of a matrix is stored on a parallel machine by representing each extended diagonal as a distributed vector. In other words the matrix D is stored in *DISTRIBUTED_ROW* form

An important aspect of the diagonal form is that it can be constructed from the simple *CONTIGUOUS_COLUMN* or *DISTRIBUTED_ROW* form without any

$$
A = \begin{bmatrix} 00 & 01 & 02 & 03 & 04 & 05 \\ 10 & 11 & 12 & 13 & 14 & 15 \\ 20 & 21 & 22 & 23 & 24 & 25 \\ 30 & 31 & 32 & 33 & 34 & 35 \\ 40 & 41 & 42 & 43 & 44 & 45 \\ 50 & 51 & 52 & 53 & 54 & 55 \end{bmatrix} \qquad D = \begin{bmatrix} 00 & 11 & 22 & 33 & 44 & 55 \\ 50 & 01 & 12 & 23 & 34 & 45 \\ 40 & 51 & 02 & 13 & 24 & 35 \\ 30 & 41 & 52 & 03 & 14 & 25 \\ 20 & 31 & 42 & 53 & 04 & 15 \\ 10 & 21 & 32 & 43 & 54 & 05 \end{bmatrix}
$$

Figure 10   Conversion of a matrix A to diagonal form D

communication To see this, note that from the expression above for the elements of D, the j-th column of the diagonal form is a permutation of the j-th column of the matrix A. Since D is stored in DISTRIBUTED_ROW form, each column of D is entirely within one processor (see section 7.1). As seen previously, a CONTIGUOUS_COLUMN or DISTRIBUTED_ROW matrix also has its columns local to processors It follows that if A is in DISTRIBUTED_ROW form, conversion of A to the DIAGONAL form D involves only in-column permutations internal to each processor.

7.3. The Areal Form of a Matrix

The AREAL form of a matrix is used to distribute a rectangular $N_x \times N_y$ matrix in a block-rectangular form to processors assigned to a logical rectangular $p_x \times p_y$ grid. Each block is surrounded by a dummy boundary of width b rows or columns The boundary allows the matrix to be treated as if stored in a global memory, provided operations on it have a stencil of radius at most b. Figure 11 displays a typical areal matrix distribution in which the sub-matrices are chosen to be square and are surrounded by boundaries of width 1 Prior to allocating matrices of type AREAL the logical grid dimensions and the boundary width must be specified by a call to the library routine *areal_parameters*():

$$areal\_parameters(p_x, p_y, per_x, per_y, b)$$

This routine constructs the logical grid mapping into the hypercube and stores

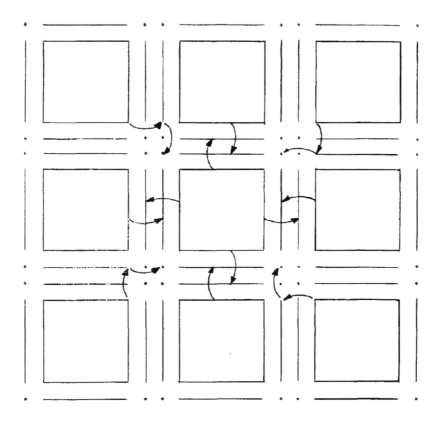

Fig. 11: The Shuffle Operation on an Areally Distributed Grid.

the resulting connectivity information and boundary width b The routine returns an error value if the logical grid cannot be mapped to the hypercube using only nearest neighbor connections. The grid is periodic in the x or y direction according as per_x or per_y is non-zero The mapping of a logical grid to the hypercube network has been discussed in section 4.1. Subsequent calls to *allocate_matrix*() with type AREAL use the information stored during the last *areal_parameters*() call, and this information is copied and stored in the matrix data structure returned by the allocation call

For simplicity we assume in this section that we can factor the number of processors as $P = p_x p_y$ and that N_x and N_y are multiples of p_x and p_y respectively: $N_x = n_x p_x$, $N_y = n_y p_y$. The processors may be labeled by a grid index as P_g, $g = (i,j)$ with $0 \le i < p_y$, $0 \le j < p_x$ The required matrix A is then conceptually split into P equal sized rectangular $n_x \times n_y$ sub-matrices A_g, with each assigned to a separate processor Matrix element $a_{i,j}$ is assigned to processor $P_{i/n_x, j/n_x}$ The *allocate_matrix*() routine allocates storage for these arrays in the processors along with storage for the boundary. Each sub-matrix is centered in an allocated matrix of size $(n_x + 2b) \times (n_y + 2b)$ Thus the block A_g is indexed as $A_{g,i,j}$ where $-b \le i < n_y + b$, and $-b \le j < n_x + b$. The rows and columns in the ranges $[-b, -1]$ and $[n_x, n_x + b - 1]$ or $[n_y, n_y + b - 1]$ correspond to the dummy boundary rows or columns.

The primary purpose of the boundary points is to allow copies of some matrix elements stored on neighboring processors to be available locally The *shuffle*() library operation:

$$shuffle(matrix, width, sides) ,$$

updates the boundary data for a sub-matrix by exchanging boundary rows or columns with all neighboring processors Only the outer *width* boundary rows or columns are exchanged, although generally *width* will be chosen to coincide with the boundary width b of the matrix The *sides* parameter specifies which of the 4 sides of the rectangles to exchange Four symbolic constants NORTH, SOUTH, EAST, WEST are used, with values 1, 2, 4 and 8 Any set of sides may be specified by orring together the corresponding symbolic constants

An important point to note is that the shuffle operation on the EAST or WEST sides involves sending non-contiguous data, assuming the sub-matrices are stored by rows Sending the individual boundary elements is too expensive on the Intel iPSC and so these elements are first copied to a buffer of appropriate size before sending. Similarly *shuffle* receives buffered sides from neighbors, then copies the data to the appropriate non-contiguous boundary locations. We have observed that the cost of these copy operations can be a substantial fraction of the communication time - see for example Table 3.

Applications of the areal distribution of a matrix will be encountered in sections 9 and 11. In these applications almost all communication is handled by the *shuffle* () operation. An important issue is the aspect ratio of the submatrices A_g allocated to each processor. Figure 12 represents mappings of logical 8×4, 16×2 and 32×1 grids onto a 32-node hypercube, and their effect on the aspect ratio of sub-matrices of a distributed matrix. In principle, the shuffle operation is proportional in cost to the perimeter of a subgrid, and can be minimized for fixed sub-grid area by ensuring that the subgrids are near square. We remark that on the Intel iPSC, where short messages are as expensive as those of length 1024 bytes, the shuffle operation is actually a fixed (large) cost operation. The largest real square sub-matrix that can be stored in an iPSC processor has about 64K elements (256K bytes) and consequently sides of length 256 elements or 1024 bytes. Thus a shuffle operation on one side consumes exactly one send and receive communication call, no matter how small the matrices are. This seriously affects the behavior of many algorithms, and also makes it essentially impossible to measure area/perimeter effects. On the Caltech Hypercube on the other hand such area/perimeter effects are readily observed, see for example Figure 26. We summarize the overheads for *shuffle* communication in

Theorem 4: The *shuffle* operation on a matrix of dimension $N_x \times N_y$, distributed areally on a $p_x \times p_y$ two dimensional mesh of processors takes time:

$$T_{shuffle} \leq 2\beta(N_x/p_x + N_y/p_y) + 2\gamma N_y/p_y + 4\alpha,$$

where the γ term represents the cost of copying non-contiguous boundary columns.

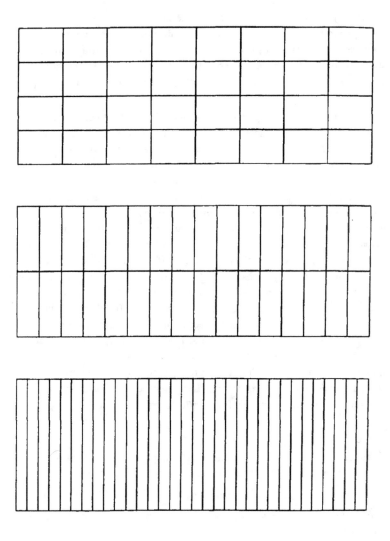

Fig. 12: Assignment of equal area grid blocks to 32 processors but with different area/perimeter ratios.

7.4. Matrix-Vector Multiplication.

In the case of the distributed-row matrix format, matrix-vector multiplication is essentially trivial In this case the multiplication reduces to a set of
standard inner products of distributed vectors Some communication may be
saved by defining processor r to be the root of the inner product tree (rather
than processor 0) when multiplying row r of the matrix with the vector - this
saves the broadcast of the result to many processors that don't really need it
The time required for the matrix-vector operation is $O(N^2/P + N\log P)$. This
algorithm may be extended to provide a matrix multiplication algorithm for
distributed matrices The row representation is only sensible for matrices
that are more or less full, so we move on now to a discussion that is
appropriate also for many sparse matrices

We will show that the diagonal representation of a matrix, combined
with the $shift\_vector()$ operation, allows the matrix multiplication to be performed efficiently and in a highly portable fashion. To introduce the algorithm, we consider first the case of a dense matrix A where we wish to compute the product $Y = AX$.

$$Y_i = \sum_{j=0}^{N-1} A_{i,j} X_j$$

$$= \sum_{j=0}^{N-1} D_{(j-i)\bmod N, j} X_j$$

$$= \sum_{m=0}^{N-1} D_{m,(i+m)\bmod N} X_{(i+m)\bmod N}$$

Denote by D_m the m-th diagonal of A in diagonal form, by V^k the vector
obtained by shifting a vector V k times, and by $\underline{*}$ the vector product of two
vectors $(u \underline{*} v)_i = u_i v_i$ Then the above equations can be written in vector
form as

$$Y = \sum_{m=0}^{N-1} D_m^m \underline{*} X^m$$

In the mth term of this sum both D_m and X are shifted m times. Instead of
shifting both the diagonals and the vector X by the same amount, we can
more efficiently shift the result vector Y by m in the opposite direction. This
leads to the final form of the matrix-vector algorithm.

$$Y = 0$$
$$\text{for } m = 0, \quad ,N-1$$
$$\quad Y = Y + D_m \underset{-}{*} X$$
$$\quad Y = Y^{-1}$$
$$\text{end}$$

The first line in the loop of this procedure is nothing but the *add_vector_times_vector_to_vector* procedure, the second line is the *shift_vector* procedure. Parallelization of both of these operations has already been discussed above Thus in order to parallelize this procedure it suffices to store the extended diagonals, D_m, X and Y as distributed vectors The computation time of the matrix-vector operation consists of $O(N^2/P)$ arithmetic operations, as is usual for a full matrix, and the time for N shift operations on the vector Y, which is $O(N)$. Communications overhead - the time spent in shifts - will be negligible compared to that in computation for large matrices, defined as those with $N \gg P$.

This algorithm is perfectly acceptable for full matrices. In the case of a sparse matrix with fill-in along diagonals, such as occurs frequently in PDE discretizations, considerable savings can be accomplished. One obvious saving is that it is unnecessary to multiply with zero diagonals. However the right hand side still must be shifted once for every diagonal. This results in savings on the number of arithmetic operations. The communication cost however remains N shift operations, although many of these can now be grouped together into longer shifts

If the bandwidth b of the matrix A is limited, we can save on the communication cost by reducing the number of necessary shifts from N to $3b$. Matrices with bandwidth b are characterized in their diagonal forms by having a cluster of extended diagonals with indices in the range $0, \ldots ,b$ and a cluster of diagonals with indices in the range $N-b, \quad ,N-1$. In this case we split the loop of vector operations describing the matrix times vector operation into two parts in the order $m=N-b, \ldots ,N-1$ followed by $m=0, \ldots ,b$ We start by initializing Y to zero, which we regard as implicitly shifted $N-b-1$ times The first loop requires $b-1$ shifts, the second b shifts. Finally, in order to get the right hand side to its correct (unshifted) place we need to shift b times in the opposite direction. Thus we obtain the algorithm

$$Y = 0$$
$$\text{for } m = N-b, \quad ,N-1$$
$$\quad Y = Y + D_m \overset{*}{-} X$$
$$\quad Y = Y^{-1}$$
$$\text{end}$$
$$\text{for } m = 0, \quad ,b$$
$$\quad Y = Y + D_m \overset{*}{-} X$$
$$\quad Y = Y^{-1}$$
$$\text{end}$$
$$Y = Y^b$$

This brings the total communication cost to $3b-1$ shifts which is a very substantial saving over the full matrix case.

As an example, in typical planar PDE discretizations over an $M \times M$ grid the resulting matrices are of size $M^2 \times M^2$ so that $N=M^2$. Usually there are $O(1)$ non-zero diagonals and the band-width b is $O(M)$. Thus time spent on communication will be $O(M)$, against time $O(M^2/P)$ spent on computation. For large problems, defined as those with $M \gg P$, communication time will be dominated by computation time.

Extending the analysis above to include communication startup costs we obtain:

Theorem 5: On a P-processor hypercube, matrix-vector multiplication for a matrix of dimension N, with bandwidth b and with d non-zero diagonals, requires time

$$T_{matrix\_vector} \leq 2d\gamma N/P + \alpha d + 3\beta b.$$

In case the matrix is symmetric some savings can obviously be accomplished in memory required. However, this will result in increased communication since the diagonals have to be shifted as well as the right hand side in one part of the calculation.

7.5. Parallel Matrix Transpose

The Matrix Transpose operation allows parallelization in a portable and efficient way for many two dimensional numerical methods, see sections 8, 9 and 14 The goal of this operation is to convert a row distributed matrix into a column distributed matrix We present two transpose algorithms and their implementations. The first algorithm uses the diagonal form of the matrices introduced above and the vector shift operation For this algorithm to be efficient, the optimized vector shift algorithm which we discussed previously is crucial. The second algorithm we call the block transpose method. Both algorithms presented require a communication time proportional to $N^2 \log P/P$ floating point number transfers, where N is the dimension of the matrix and $P = 2^D$ is the size of the hypercube. These estimates are based on the assumption that the cost of communications is linear in the amount of data transferred As discuss previously this assumption is not valid for the iPSC In particular the transpose based on the shift algorithm is strongly affected by the startup cost for communication, and consequently appears much slower on the iPSC than the block transpose.

7.6. The Shift Transpose Algorithm

There are two key points to understanding this transpose algorithm. First, it is easy, and requires no communication between processors, to convert a matrix from the row distributed form to the diagonal form This issue has already been addressed in section 7 2 where the diagonal form was introduced. Secondly, it is straightforward to transpose the diagonal form. We now examine the second point in greater detail.

The element $A_{i,j}$ of the matrix A is equivalent to element $D_{(j-i) \bmod N, j}$ of the diagonal form D. The transpose A^T of the matrix A, has a diagonal form

DT such that

$$DT_{(j-l)\bmod N,j} = A^T_{i,j} = A_{j,i} = D_{(i-j)\bmod N,i}$$

From this it follows that

$$DT_{k,j} = D_{N-k,i} = D_{N-k,(j-k)\bmod N}.$$

This means that the kth diagonal of DT is the $(n-k)^{th}$ diagonal of D, shifted by k. Note that once $k > N/2$ it is cheaper and equivalent to shift by $N-k$. As a result, the cost of this algorithm, using the straightforward shift operation, is $N(N-1)/4$ element transfers, which makes it comparable in cost to transposing the whole matrix on a serial machine. Thus this algorithm makes no effective use of parallelism[1] Using the optimized shift operation, the number of communication operations is reduced asymptotically to $(\log P - 1)N^2/P$, which, apart from the $\log P$ factor, indicates almost optimal use of parallelism. For a 5D cube this is about a factor of 2 better than using the straight shift while for a 7D cube the gain is about a factor of 5.3 A much more detailed analysis of this algorithm is presented in the following subsection.

In the implementation of this transpose algorithm for a matrix A of *DISTRIBUTED_ROW* type, we never actually store the matrix D Instead of storing the whole diagonal form, we allocate two distributed vectors of type *SHIFT*. At the kth step, extended diagonals k and $N-k$ of A are constructed in these vectors and respectively shifted by k and $-k$ Once these shift operations are performed, the diagonals are copied back to the matrix, taking care that the diagonals are interchanged correctly

7.6.1. Communication Cost for Shift Transpose

Theorem 6: The Shift Transpose of an N by N matrix on a P-processor hypercube takes time:

$$T_{shift\_transpose} \leq (\beta \log P + \gamma)N^2/P + \alpha N \log P$$

Proof:

The communication cost for transposing an $N \times N$ matrix using this algorithm is incurred by shifting the m-th (extended) diagonal m times, with m taking on the values $0, \cdots, N-1$. For each m, we can find m_s and m_e such that $m = m_s h + m_e$ $(0 \le m_e < h)$, where $h = P/N$. The m element shifts required for diagonal m can be obtained by performing m_s segment shifts and m_e element shifts. As m ranges from 0 through $N-1$, m_e assumes each of the values $0, \quad , h-1$ P times and m_s assumes each of the values $0, \quad \cdot , P-1$ h times.

From the range of values assumed by m_e, we can determine the total cost T_e associated with the element shifts:

$$T_e = P \sum_{k=1}^{h-1} ST(k) ,$$

where we have used notation introduced in section 2 3 $ST(k)$ is the cost of communicating a segment of length k to a neighbor processor.

We focus our attention now on the segment shifts Since segment shifts over more than $P/2$ processors are better done as segment shifts in the opposite direction, we can divide the set of values assumed by m_s into three subsets: $[1, P/2-1]$, $[-P/2+1, -1]$ and the single value $P/2$ As indicated earlier in discussing the optimized shift algorithm, the actual number of segment transfers to be performed to complete a segment shift over m_s processors is most easily determined by considering set bits in the binary representation of m_s: bit 0 represents 1 segment transfer, the other bits represent 2 physical segment transfers Since $P = 2^D$, the values in the first range are represented by $D-1$ bits and as the range is traversed each bit is set half of the time, i.e. in $P/4$ cases. It follows that the total number of physical segment transfers for the first range of values of m_s is: $(2D-3)P/4$. The second range of m_s values requires by symmetry an identical number of segment transfers, while the final value $m_s = P/2$ requires 2 segment transfers Since each value m_s is assumed h times and the cost of a segment transfer is $ST(h)$ it follows that the total cost of all segment shifts required in the transpose algorithm is·

$$T_s = ((D-1.5)P+2) \, hST(h).$$

Adding together the quantities T_e and T_s derived above, and using the relation $ST(k) = \alpha + \beta k$ introduced in section 2.3, we obtain as the total time for the shift transpose of a full matrix

$$T_{shift} = ((D-1.5)P+2) \, h \, (\alpha+\beta h) + P(h-1)(\alpha+0.5\beta h)$$
$$\leq \beta N^2 \log P \, /P + \alpha N \log P$$

The term proportional to α gives the effect of message startup costs and can dominate for small N on machines such as the iPSC where $\beta \ll \alpha$. To complete the proof of the theorem we add in the computation cost related to the copying of data internal to processors.

The shift transpose algorithm is well-suited to diagonally sparse matrices, where as in the matrix-vector multiplication algorithm only the non-zero diagonals actually require shifting.

7.7. The Recursive Block Transpose Algorithm

An algorithm for transposing a matrix in a time of order $N^2 \log P/P$ is possible without requiring that the size of the matrix be a multiple of the number of processors (as was necessary for the optimized shift operations used in the previous section) Note that a 2×2 matrix is transposed by exchanging its upper right and lower left elements This leads to the following serial algorithm to transpose a matrix

$$Transpose(M):$$

$$M = \begin{bmatrix} M_{00} & M_{01} \\ M_{10} & M_{11} \end{bmatrix}$$

$$M_{01} \longleftrightarrow M_{10}$$

$$Transpose(M_{00}) \qquad Transpose(M_{01})$$
$$Transpose(M_{10}) \qquad Transpose(M_{11})$$

A matrix with dimension $N = 2^L$ can be transposed by recursive application of this procedure through L levels. At the final level all matrices are 1×1 and so are trivially transposed We indicate the first three levels in this procedure in Figure 13, where the arrows represent block interchanges.

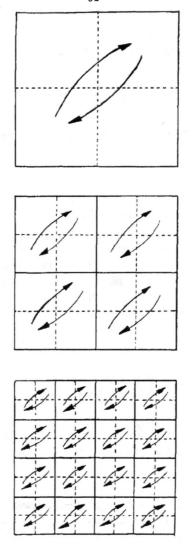

Fig. 13: Schematic representation of the Block Transpose algorithm as it would be executed on a 2 Dimensional Hypercube (4

To implement this algorithm on a hypercube we store the matrix M in contiguous-row form and assume that its dimension N is divisible by the processor number $P = 2^D$: thus $N = h2^D$. The case where the size of the matrix is not a multiple of the number of processors does not pose significant problems. For this algorithm we order the processors according to natural ordering. Thus processors that are a distance 2^d apart in processor number differ in only one bit in their binary representation and are therefore nearest neighbors on the network. It follows that at level L in the algorithm the matrices are of dimension $h2^{D-L}$ and are stored as contiguous-row distributed matrices over 2^{D-L} processors. Furthermore these processors are a contiguous sequence of nearest-neighbor processors and form a $D-L$ dimensional sub-cube of the hypercube. At level D all sub-matrices are local to one processor and may then be transposed by the normal serial algorithm. Thus we terminate the recursion at level D.

The crucial point in analyzing this algorithm is that the interchange of the upper right and lower left sub-block at level L involves communication between processors with processor numbers differing by 2^{D-L-1}, hence the two communicating processors are at a physical distance of 1. In principle the transpose of the 4 sub-blocks can be done in parallel. However since the 4 blocks are distributed over only 2 subcubes, the effective parallelism is only 2. Equivalently, the two upper blocks span the same set of processors; the two lower blocks also span a common set of processors

The total time for this algorithm is asymptotically $O(N^2 \log P/P)$ as shown in the detail in the following subsection.

7.7.1. Communication Cost for the Block Transpose Algorithm

Theorem 7: The Block Transpose of an N by N matrix on a P-processor hypercube takes time:

$$T_{block\_transpose} \leq ((\alpha/\lambda + \beta + \gamma)\log P + 1.5\gamma) \, N^2/P + 2(\alpha + \beta\gamma)P.$$

Proof:

The fundamental communication taking place in the block transpose algorithm, is the transfer of a block of k rows of length l (the submatrices) from one processor to another. The cost associated with this transfer, which we denote by $BT(k,l)$, depends on the implementation details of the transfer, and we consider three possibilities. The most efficient solution is to send a contiguous array of kl elements, entailing a block transfer cost of·

$$BT^1(k,l) = ST(kl).$$

where $ST(k)$ is the segment transfer cost introduced in section 2.3. The rows of the individual submatrices are not stored contiguously however. This makes it more convenient to send over individual rows of the matrix one by one, in which case we obtain

$$BT^2(k,l) = kST(l).$$

Because of the high cost of sending short messages, this procedure is very expensive on the iPSC. We avoid this by using a buffering scheme to collect the non contiguous rows before transmission. In practice this is accomplished by calling the *send_local_vectors*() routine from the vector library. With a limited buffer length λ, and including time necessary to copy the data into the buffer, the estimated communication time is.

$$BT^3(k,l) = (kl)/\lambda\ ST(\lambda) + \sigma ST((kl)\bmod \lambda) + \gamma kl\ ,$$

where in the first term integer division by λ is assumed and σ is 0 if kl is a multiple of λ and is 1 otherwise.

Let B_d be the *communication* time required to block transpose a matrix of dimension $2^d h \times 2^d h$ on a d dimensional hypercube, which we call a level d transpose. As shown above, the level d transpose is reduced to 4 level $(d-1)$ transpose operations and a submatrix exchange. The effective parallelism in the transpose operations is 2, in the exchange it is 2^{d-1}. Using the block transfer time introduced above, we obtain the following recursion relation:

$$B_d = 2B_{d-1} + 2BT(h, 2^{d-1}h)\ ,$$

with an initial value $B_0 = 0$. It follows that:

$$B_D = P \sum_{i=0}^{D-1} 2^{-i} BT(h, 2^i h)\ .$$

We can now substitute the appropriate expression for BT^i in terms of ST and use the relation $ST(k) = \alpha + \beta k$ from section 2.3. We obtain as bounds for communication cost in transposing an $N \times N$ matrix, where $N = Ph = 2^D h$:

$$B_b^1 \leq 2\alpha P + \beta N^2 \log P/P,$$
$$B_b^2 \leq 2\alpha N + \beta N^2 \log P/P,$$
$$B_b^3 \leq 2(\alpha + \beta\lambda)P + (\alpha/\lambda + \beta + \gamma) N^2 \log P/P.$$

We notice that for asymptotically large N, the buffered mechanism is the most expensive as expected since for large enough N the individual rows exceed the buffer length. Since the first method is not practical due to the matrix storage scheme, the second method is therefore indicated for large matrices. For smaller matrices on a machine like the iPSC the third method has a great advantage because of the smaller coefficient of the term proportional to α, the message startup cost. In fact using actual iPSC measurements we find that for all but the largest matrices that will fit on the hardware, the term proportional to α plays a dominant role, making the buffered scheme a necessity. To complete the proof of the theorem, we add the computation cost involved in performing P serial transposes of $N/P \times N/P$ matrices in each processor, which consumes time $3/2\gamma N(N/P-1)$

7.8. A Comparison of Transpose Algorithms on the iPSC.

We now compare the best Block Transpose method with the Shift Transpose. From the higher order terms in N it would seem that both methods are equivalent as far as communication time is concerned. The linear term in the Shift Transpose cost however dominates for practical problems on the iPSC. Even for the largest matrix that can be stored on the iPSC (2048×2048), the segment length of the distributed vectors used in the algorithm is only 16. This means that all communications taking place are well under the effective minimum message length of 1024 bytes or 256 floating point variables. For the Block Transpose, the number of floating point variables per message is proportional to N^2/P^2. For the largest problem this is 256 elements per smallest submatrix encountered. Therefore only the values of α and β for long messages play a role if the buffering scheme described above is used. For smaller problems, the short message values influence timings in the exchange of the smaller submatrices. These remarks are borne out by the coefficient of the term proportional to α in Theorems 6 and 7. In the Shift Transpose case this coefficient is proportional to N while in the Block Transpose case it is proportional to P and is in fact therefore independent of matrix size.

For clarity, we have neglected the "computation" cost in some formulae above, although this is included in the theorem statements. The basic computation unit for the transpose algorithms is the assignment of an indexed floating point variable, i.e. a statement of the form: $x = y_i$. Assuming this statement takes time γ, we have expressed the computation time as a multiple of γ in Theorems 6 and 7. For the algorithms studied, executed on the iPSC, these costs turn out to be negligible. For example we find that the "computation" cost on a 128 node iPSC for a 2048×2048 matrix is about 10% of the measured transpose time. Thus the transpose is dominated by communication

7.9. Results and Timings

In the figures for this section we always display times as a function of matrix *size* i.e. of the number of elements N^2. On a serial machine the cost of matrix transpose is a linear function of its size. In Figure 14 we display the cost of transposing a matrix on the 5D Caltech Hypercube as a function of the size of the matrix. The timings displayed, use the transpose algorithm based on the diagonal form and the non-optimized vector shift operation.

Figure 15 presents corresponding timings for the optimized shift transpose algorithm and the block transpose algorithm as a function of matrix size N^2 on the 128 processor iPSC. Note that, for the largest matrix that can be stored on the iPSC (dimension 2048) the block transpose is much faster than the shift transpose, despite the fact that the latter is asymptotically slightly faster. The linear form of the block transpose curve indicates that that algorithm has already attained its asymptotic behavior. In Figure 16 we display the timing for the block transpose on a better scale where it becomes clear that the curve has in fact approached its asymptotic rate after an initial high startup overhead for small matrices.

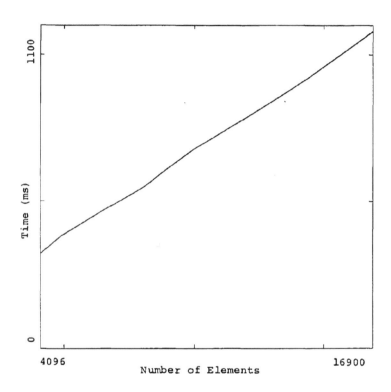

Fig. 14: Cost of Transposing a matrix on the 32 node Caltech Hypercube as a function of the Number of Elements. Transposing using the Non Optimized Vector Shift algorithm.

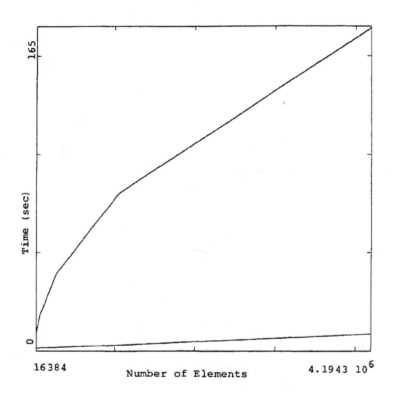

16384 Number of Elements 4.1943 10^6

Fig. 15: Time to Transpose a matrix with the Shift Transpose (upper curve) and the Block Transpose (lower curve) algorithm respectively on the 128 node iPSC as a function of the Number of Elements.

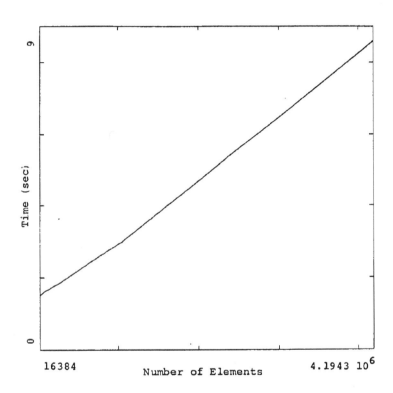

Fig. 16: Time to Transpose a matrix with the the Block Transpose algorithm on the 128 node iPSC as a function of the Number of Elements.

8. TRANSPOSE SPLITTING ALGORITHMS

We introduce here a general class of algorithms which may be solved on parallel machines with almost no modification to an appropriate serial program. We term these algorithms *Transpose Splitting Algorithms* because the application of a grid-point operation is split into two parts - one applied to the x-direction and one to the y-direction. Examples are operator-splitting hyperbolic methods and ADI methods, though many other algorithms may be recognized to be of this type.

Consider a two-dimensional function $M(x,y)$ on a grid, where x,y label the rows and columns of grid-points. We store the values of M as a contiguous-row distributed matrix (see Section 7 1), with one or more complete rows of points per processor Many numerical algorithms amount to applying one or more operations $O(x,y)$ successively to M at each point x,y, which we denote by the notation $O() \cdot M()$. This would be represented in code as a loop of the form·

$$\text{for } each \; x,y: \; O(x,y) \quad M().$$

Suppose that the operation O may be split into x and y parts which we represent schematically as:

$$O(x,y) = B(y) \cdot A(x), \; x,y \; in \; grid$$

Now assume that $A(x) \cdot M(x,y)$ applied at point x,y uses only values $M(x,y)$ from the same row y containing the point. Similarly we assume that $B(y)$ applied at x,y uses only values $M(x,y)$ from the same column x. An equivalent code segment for application of O would then be.

$$\text{for } each \; y$$
$$\quad \text{for } each \; x: \; A(x) \quad M()$$

$$\text{for } each \; x$$
$$\quad \text{for } each \; y \quad B(y) \cdot AM()$$

Note than in the application of A the inner loop is along a row, and since $A()$ involves only elements from the same row, and since each row is entirely in one processor, it follows that the inner loop involves no communication whatever.   The outer loop, the loop over rows, also involves no communication and is furthermore completely parallelized. Thus A may be applied to M without communication and in an optimally parallel fashion.

Unfortunately the loop with B is not so nice. Here there is communication every time that B involves adjacent column elements that happen to be in different processors. However there is a trivial trick that removes all communication from the B loop and also parallelizes it. The idea here is first to apply a transpose to M, or rather to the matrix $A\,M$. Then the roles of x,y in the B loop are reversed and the relevant code becomes:

$$N = transpose(A \cdot M)$$

for *each* y
   for *each* x: $B(x) \cdot N()$

All communication between processors in the application of O to M is now hidden in the transpose operation. Outside of this the code is optimally parallel.

Even for situations in which $A(x)$ applied at point x,y uses values of $M(x,y)$ from another row, this algorithm can still provide a useful approach to parallelism. To accomplish this one stores the matrix M in AREAL form but in the extreme case where the processors are assigned as a $1 \times P$ matrix. Then a vertical *shuffle* is used to ensure that the data required by $A(x)$ when operating on boundary rows of sub-grids are available as needed. After a transpose and *shuffle* the data required for B will also be available as needed

9. HYPERBOLIC EQUATIONS

The hyperbolic equations we have studied are of the form of conservation laws:

$$\frac{\partial u}{\partial t} = - \nabla \ F\left(u\right) .$$

The nonlinearity of the function $F(u)$ may result in the formation of discontinuities (shocks) in a finite time. Hence, the numerical methods used to solve such equations must also be able to handle discontinuities of the solution. The Random Choice Method[16] is one solution method that is sensitive to shocks We have also implemented simple finite-difference schemes such as the Lax-Wendroff method.[17]

We have parallelized the hyperbolic solution in several different ways. In one approach, the parallelization of these methods is a typical application of the Transpose Splitting strategy, which we have discussed as a general method in section 8. The application to hyperbolic equations will be given in section 9.3. Another approach is based on decomposition of the rectangular grids into rectangular subgrids of near-square aspect ratio, with each subgrid assigned to a separate processor, see sections 7.3 and 9 2 We have implemented the first approach on the Caltech Hypercube We compare both methods on the Intel iPSC and conclude that the former involves substantially less development effort, while yielding comparable efficiency to the latter approach

Our test problems have been model hyperbolic equations - scalar equations with simple non-linearities Real physical problems involve complex equations of state and may require substantial amounts of numerical computation per grid point. Real problems typically involve only the same communication costs as the model problems however, unless they are discretized using very high-order schemes. To assess the behavior of hypercubes on such codes, we use the same hyperbolic solvers mentioned above, but with a dummy non-linear function that simply executes n floating point operations. We preserve all communication operations exactly as they would be in the hyperbolic solver This allows us to measure efficiency as a function of work done per grid-point, and of the number of grid-points The results of these studies are very favorable. If significant amounts of work are done per grid-point, hypercubes can provide near optimal efficiency even on very coarse grids, down to a few dozen points per processor The crucial point here is that the model problems already perform well. This ensures that real problems will perform even more successfully In these studies we use the *computational efficiency* of a program, defined as the ratio of computation time to execution time of the program, as a measure of communication performance These results are discussed in detail in section 10, although the methods used to parallelize the programs are developed in this section

9.1. Discretization Methods

We have developed parallel hyperbolic equation solvers based on both finite difference and random choice methods. While standard difference schemes vectorize easily, this may not be so for the random choice method since different random sequences may be used in different spatial locations and because the underlying Riemann solutions generally involve complex logic However random choice parallelizes as easily as do difference methods. In addition, realistic physical problems involve complex equations of state which may be difficult to vectorize, but again these parallelize readily. These are examples where MIMD parallelism has a real advantage over SIMD parallelism.

The random choice method (RCM in the sequel) is a general purpose solution method for hyperbolic conservation laws, allowing formation of shocks without affecting convergence of the method. A fundamental building block of the method is the one dimensional Riemann solver for the conservation law under consideration. Given a step function at time t describing two constant states connected with each other through a shock, the Riemann solver returns the solution for the given conservation law at time $t + \Delta t$.

The RCM uses a Riemann solver to construct approximate solutions of one-dimensional conservation laws for all time. The function describing the state at time t is approximated by a sequence of step functions. At each jump of the step function a Riemann problem is solved, resulting in a solution that is no longer piece-wise constant. The process is then repeated. To extract a step-function from an approximate solution, a random point is chosen in each of the elementary Riemann solutions. It can be shown[16, 18] that this method converges to the correct solution even in the case that shocks are present in the system In two dimensions, the RCM can still be applied using operator splitting, first performing one-dimensional random choice solutions along each row, and then along each column.

In parallelizing this method we have distributed the state variables $u(x,y)$ in several ways. We have represented the solution as both an areal-distributed grid function, using a distributed matrix of type $AREAL$, and as as a row-distributed grid function, using a distributed matrix of type $CONTIGUOUS\_ROW$ - see section 7 1 and 7.3 for discussion of these matrix types. The communication pattern varies with the distribution scheme and we analyze the communication costs of each scheme in the following subsections. In the areal case communication is provided by the $shuffle()$ operation whereas for the row case the $transpose()$ operation is used. There is one further issue involved in the parallelization. As is normal for explicit hyperbolic solution methods, the length of the time-step Δt is constrained by a Courant condition based on the maximum wave-speed The computation of the maximum wave-speed uses the max_vector routine to compute the global maximum of the local processor maxima and broadcast the results back to the individual processors Because of the semantics of the max_vector routine, all processors can compute the new time step independently of each other, although of course they obtain the same result. As a result the algorithm is completely independent of node location in the network With both distribution methods all communication is handled entirely by the vector and matrix libraries.

The same parallelization schemes can be applied to finite difference schemes In addition to Random Choice we have used the Lax-Wendroff method.[17] The key point is that both methods involve only local stencils of a few grid points. For the row-distribution method it is also important that the discretization scheme be amenable to operator splitting.

9.2. Areal Decomposition

In the areal decomposition, a logical $p_x \times p_y$ rectangular grid is mapped onto the hypercube network as discussed in section 4.1. Thus the processors may be labeled by a grid index as p_g, $g=(i,j)$ with $0 \le i < p_y$, $0 \le j < p_x$. For simplicity we assume in this section that the we can factor the number of processors as $P = p_x p_y$ and that the computational domain is a rectangular $N_x \times N_y$ grid where $N_x = p_x n_x$ and $N_y = p_y n_y$. A grid function $U = \{U_g\}$ representing the solution values of the hyperbolic equation on the $N_x \times N_y$ grid is now allocated as a distributed matrix of AREAL type with boundary width 1 as described in section 7.1 and 7.3. Thus the component U_g is indexed as $U_{g,i,j}$ where $-1 \leq i \leq n_y$, $-1 \leq j \leq n_x$, with the rows and columns numbered -1 and n_x or n_y being the extra boundary rows To perform the hyperbolic time step computation on the grid, a *shuffle* operation, see section 7.3, is first performed in order to fill the boundary perimeter of each U_g with the appropriate values from the neighboring subgrids. Following this, the standard sequential hyperbolic method is performed in the interior of each subgrid, i.e. for rows and columns in the range $[0, n_y - 1] \times [0, n_x - 1]$ In the course of this computation the maximum wave speed within that subgrid is computed, and is stored into the local component of a distributed vector of length P. The final phase in the solution is the computation of a new time step, which is based on a Courant condition involving the maximum wave-speed To compute the maximum wave-speed, the *max_vector* operation is applied to the vector of local maxima, which provides the global maximum value to all of the processors - see section 6.5 for the semantics of *inner_product* and *max_vector*. This allows all of the processors to compute the next time step simultaneously, using only vector/matrix library routines for all communication activities, and ensuring an SIMD form for the computation in cases where the local discretization operation is translation independent

9.2.1. Communication Costs:

For simplicity of analysis, we begin by assuming that the number of processors is a perfect square, $P = p^2$, and that the computational domain is a square grid of N points, with $N = n^2 p^2$. Thus there are n^2 grid points per processor For any hyperbolic scheme, the computation to be performed per grid-point is a local operation in grid-space and thus we regard the basic computation per grid-point as an $O(1)$ operation. We stress that it may in fact be a large number if the physics or the discretization method is complex. Consequently the cost of the hyperbolic sweep over a sub-grid is $O(n^2)$, proportional to the area of the subgrid. Since all sub-grids are computed in parallel, this is also the cost of the sweep over the whole grid. The cost of the shuffle exchange operation at the start of the time step is $O(n)$, since the amount of data to be transferred is proportional to the perimeter of the subgrids, and the communications for different subgrids are performed in parallel. Finally the time step computation involves a cost of $O(\log P)$, which is essentially negligible since the maximum is evaluated over a tree. It follows that the computation time per time step for the complete hyperbolic solution is $O(n^2)$, i e $O(N/P)$, and that communication costs are negligible compared to computation costs provided only that sufficiently many grid points are stored in each processor, *independent* of the number of processors. Equivalently, the computational efficiency of the algorithm is given by

$$E = T_{computation}/T_{execution} = \frac{1}{1+O(1/n)}$$

If the time step had been evaluated by accumulating the maxima in a single processor, then the time step computation would be $O(P)$ and might actually dominate for problems such that $N<P^2$. We remark that machines which involve a high overhead for sending short messages, provide especially poor performance for the areal algorithm on coarse grids, since the subgrid perimeters may be smaller than the effective minimum packet size. In particular, on the Intel iPSC the effective minimum packet size is 1024 bytes, so that a subgrid edge must contain at least 256 points for fully-efficient communication. Thus a subgrid would contain 65536 grid-points, close to the maximum available user memory (about 275K bytes including the program). Below this size the above analysis is not strictly accurate, since the cost of a boundary shuffle should be taken as $O(\alpha+\beta n)$ in the notation of section 2 3, resulting in a much lower efficiency for small n when α is large as in the iPSC case. In this case the efficiency has the asymptotic form

$$E = \frac{1}{1+O(\alpha/n^2 + \beta/n)}$$

Returning to the more general case of rectangular sub-grids, we now study the behavior of the computational efficiency as a function of the aspect ratio $a=n_x/n_y$, the ratio of height to width of the subgrid, where we assume without loss that $n_x \leq n_y$. The computation cost is $O(n_x n_y)$, whereas the communication cost is $O(n_x + n_y)$, ignoring the time step cost. It follows that the computational efficiency of the algorithm is

$$E = T_{computation}/T_{execution} = \frac{1}{1 + O(1/n_x + 1/n_y)}$$

This expression is maximized for fixed sub-grid size $n_x n_y$ when $n_x = n_y$, which corresponds to the case of square sub-grids. The efficiency is minimized by allowing n_x to be as small as possible, which occurs if $p_x = P$ and $p_y = 1$ corresponding to the bottom configuration in Figure 12. This corresponds to a linear decomposition in which whole columns are distributed to each processor, so that the state variables are effectively represented as a contiguous-column matrix. Assuming that the overall grid is square so that $N_x = N_y$, it follows that $n_x = \sqrt{N/P}$, and $n_y = \sqrt{N}$ Correspondingly the minimal efficiency may then be expressed as:

$$E = \frac{1}{1 + O(P/n_x n_y)^{1/2}},$$

where we note that $n_x n_y$ is the number of grid-points per processor. The efficiency may no longer be close to 1 if the number of processors exceeds the number of grid-points per processor. In particular, for fixed processor memory size (which limits the subgrid size $n_x n_y$), the efficiency goes to 0 as larger problems are solved with correspondingly more processors

9.3. Linear Decomposition + Transpose

For this section we assume that the hyperbolic discretization method is of operator splitting type and consists of a grid sweep with a horizontal stencil operator in the x direction, followed by a grid sweep with a vertical stencil operator in the y direction. We are then in a position to apply the Transpose Splitting Parallelization discussed in section 8. Thus the state variables are distributed as a matrix in *CONTIGUOUS_ROW* form, in the notation of section 7.1. In particular, N/P rows of grid-points are assigned to each processor. The hyperbolic sweep (one-dimensional) is now applied to each row, involving no communication cost since all required stencil values are in a single row and are thus local to a processor. Following this, the matrix of state values is transposed, so that effectively it is now in *CONTIGUOUS_COLUMN* form Then the hyperbolic sweep is applied in the y-direction, which now means along rows, and thus again no communication is required Finally the time step is computed using the *max_vector* routine. In the following time step we perform the y-sweep before the x-sweep which saves on performing one transpose per time-step. This algorithm differs from the corresponding algorithm on a shared memory machine or on a serial machine only in one line - the presence of the matrix transpose. Indeed, even on a serial machine there may be an advantage in performing the transpose if there is a hardware cache present.

9.3.1. Communication Costs

As usual, the computation cost is proportional to the number of grid points per processor and is thus $O(N/P)$ The only communication costs in this version of the algorithm are the transpose and the negligible $O(\log P)$ cost of the *max_vector* routine. As discussed in section 7.7, the transpose may be performed using the block transpose algorithm in $O(N\log P/P)$ time. It appears that communication is comparable to computation on all size machines. In fact communication is always small compared to computation, because the stencil operation performed per grid-point in a non-linear conservation law is typically non-trivial - i.e. has a large value for its $O(1)$ cost compared to the corresponding $O(1)$ cost for the communication. We note that since the block transpose communicates matrix elements in long sequences (see section 7.7), this algorithm behaves well even on machines with a high overhead for communicating short messages This is borne out in the iPSC experiments reported below.

9.4. Timing and Results for Hyperbolic Equations

In Figure 17, the time used on the 32 node Caltech hypercube to perform a Random Choice Method iteration step is given as a function of the size N of the problem. The method used was the operator splitting algorithm with transpose. The lower curve in this figure, is the time needed to transpose the solution matrix The transpose algorithm in these timings used the non-optimized vector shift operation Even in this case, the communication cost (transpose) is already dominated by the computation cost

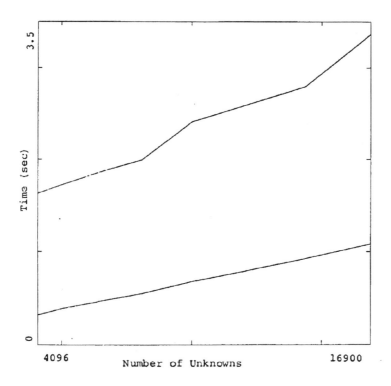

Fig. 17: Iteration Time (in ms) for the Random Choice Method on the 32 node Caltech Hypercube versus Size of the Problem. The lower curve is the time necessary to transpose the matrix (Shift Transpose with Non Optimized Vector Shift).

Analogously, Figure 18 and Figure 19 give the iteration time and the computational efficiency of the Random Choice Method as a function of the size of the problem on the 128 node iPSC. In these iPSC calculations the recursive block transpose algorithm was used instead of the shift transpose used in the Caltech calculations. In Figure 18 we note that as the grid becomes large both the computation and transpose times are converging to straight lines as expected from the previous discussion - times proportional to the number of grid points N. However the transpose time has a smaller slope, ensuring that it will always be dominated by computation time. For very small grid sizes however one can clearly see from Figure 18 that the transpose cost dominates the computation due to the high communication startup cost on the iPSC. This shows up more clearly in Figure 19 where computational efficiencies are poor for very coarse grids. For the Lax Wendroff method using either of the transpose algorithms, we obtained equivalent results.

We have also used the areal decomposition technique for each of the methods, with rectangular sub-grids ranging in shape from near square to thin strips. As an example, we display in Figure 20 the scaled time step cost for the areal-distributed Lax-Wendroff method on a 32 node iPSC as a function of the number of unknowns in the problem. Here the time per time step has been divided by the number of grid-points to arrive at a measure of effective processing time per grid-point and to allow comparison of different grid-sizes. The areal decomposition in this case corresponded to mapping the 32 processors onto an 8×4 rectangular grid The scaled time decreases rapidly to a limiting value as the grid size increases, corresponding to increasing computational efficiency. The scaled time is proportional in fact to the inverse of computational efficiency

Table 7 indicates the effect of the area-perimeter law as a function of subgrid rectangle shape on the iPSC for a fixed-size solution The 128 processors are mapped to logical rectangular $p_x \times p_y$ grids of dimensions 16×8, 32×4, 64×2 and 128×1 respectively, and the 4,096,000 unknowns are then distributed to the processors as indicated previously. As discussed above it is extremely difficult to measure the area-perimeter effect on the iPSC because startup time dominates communication. This is borne out clearly in the table. The near-square 16×8 grid is slightly faster than the following two more rectangular mappings, but the linear distribution is fastest. The explanation is

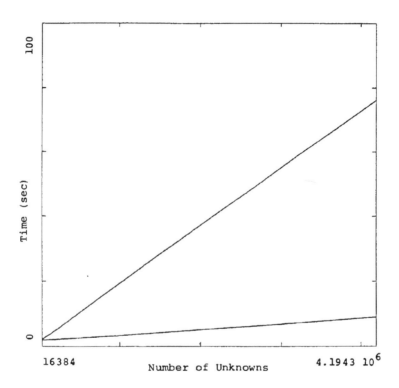

Fig. 18: Iteration Time (in ms) for the Random Choice Method on the 128 node iPSC versus Size of the Problem. The lower curve is the time spent in communication (transpose time with Block Transpose algorithm and computing of global maximum speeds for the Courant condition).

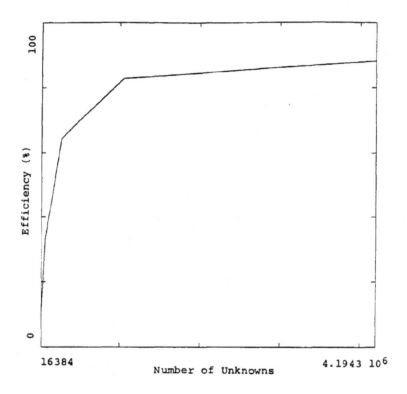

Fig. 19: Computational Efficiency of the Random Choice Method using Matrix Transpose versus Size of Problem on the 128 node iPSC.

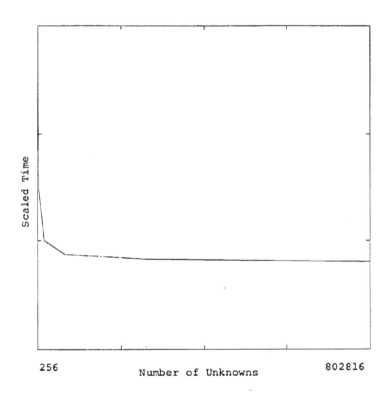

256   Number of Unknowns   802816

Fig. 20: We compare Iteration Time of the Areal Decomposition Lax Wendroff Method for different size problems on the 32 node iPSC by scaling the times by the number of floating point operations executed.

Table 7: IPSC Area/Perimeter Effect		
P_x	P_y	Time (secs/iteration)
16	8	94 576
32	4	95 248
64	2	95 376
128	1	91 968

that the linear distribution is a limiting case and involves special effects For example, no communication is required across two of the sides in a linear distribution Thus only the first three entries in the table are really relevant and the observed area-perimeter effect is small. For comparison with the Caltech Hypercube, we refer to the corresponding Figure 24 where aspect ratio effects are quite pronounced.

10. COMPARISON OF ALGORITHMS AND ARCHITECTURES

In this section we discuss the amount of computational work per grid-point that is necessary to maintain efficient use of the hypercube in typical numerical procedures. For this purpose we define the *computational efficiency E* of a program running on a parallel computer as the ratio of computation time to total execution time:

$$E = \frac{T_{computation}}{T_{execution}}$$

This ratio has the desirable property that efficiency increases if communication time can be overlapped with computation. It does not however measure efficiency relative to the best serial algorithm for the same problem

The computational efficiency of a program with localized communication will generally depend strongly on the grid size. As the number of grid-points per processor increases, so does efficiency. The efficiency will also depend on the number of floating point operations performed per grid-point - if more computation is performed, efficiency will be increased. In section 10.1 we describe measurements we have performed of efficiency as a function of work per grid-point for the hyperbolic solvers described in section 9 with both areal and transpose algorithms. Section 10.2 discusses the general applicability and utility of these measurements for the comparative evaluation of architectures and algorithms.

10.1. Efficiency and Work per Grid-point

To measure these effects, we have used the basic transpose-based hyperbolic solver described previously. We replaced the low-level two or three-point hyperbolic difference step by a call to a subroutine that executes some number ($NFLOPS$) of floating point multiplications. By running the resulting program with different grid sizes and values of $NFLOPS$ we obtain performance curves for the iPSC that measure efficiency with respect to both grid-size and work per point. We summarize these results with the curves in Figure 21, obtained on a 128 processor iPSC.

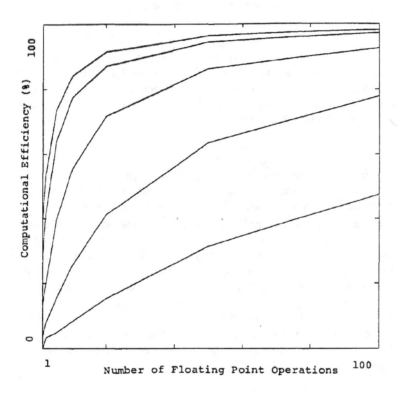

Fig. 21: Computational Efficiency on the 128 node iPSC for an Areal Sweep with Block Matrix transpose versus Number of Floating Point Operations executed per grid point. The curves from top to bottom are for grid sizes: 2048*2048, 1024*1024, 512*512, 256*256, 128*128.

We conclude that computational efficiency is excellent in almost all cases
($E \geq 50\%$). Even on very coarse grids, such as 128×128 grids where there
are only 128 grid-points per processor, reasonable efficiencies may be
achieved by executing less than a hundred operations per grid-point (In fact
we have had excellent efficiencies with even 32 points per node).

The hyperbolic equations we have solved in this paper have been model
problems, where the cost of computation at a grid point is just a few multiplications. From the graphs in Figure 21 it is clear that the solution of realistic
physical equations using real equations of state or other material properties
will be extremely efficient, even on very coarse grids We stress again that
there are three other factors that are likely to further increase computational
efficiency:

a) the current iPSC design has a very high communication overhead

b) future machines will have much larger memories, and so many more grid
points per processor

c) the transpose algorithm is far from optimal with respect to communication

As an illustration of point c) we repeated the above experiment using the
areal decomposition approach discussed in section 9 1. The resulting efficiency curves are presented in Figure 22 and are seen to be substantially
better than in the transpose case.

We conclude therefore that the vast majority of regular-grid hyperbolic
methods will parallelize with an efficiency close to 1, provided only that the
memory available on each processor is large enough to store the code, associated data and results needed by that processor for a reasonably large number
of grid-points. It is important to note that for methods such as Random
Choice where the computational work varies from point to point in a grid,
the efficiency as described above will be misleading in that some processors
may in fact be idle while others are completing their locally expensive computations. Thus unless load-balancing or time-sharing of processors is
employed, efficiencies for such methods will be lower than predicted unless
the value of *NFLOPS* used in the predictions corresponds to the worst case
possible at any grid-point

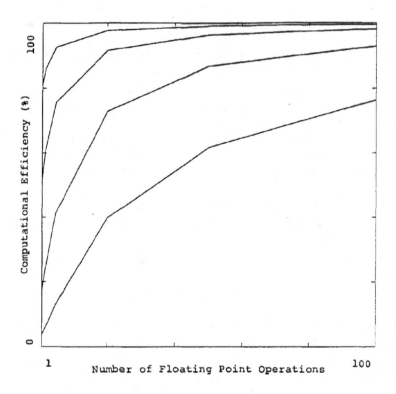

Fig. 22: Computational Efficiency on the 128 node iPSC for an Areal Sweep with Areal Decomposition Technique versus Number of Floating Point Operations executed per grid point. The curves from top to bottom are for grid sizes: 1024*1024, 512*512, 256*256, 128*128.

10.2. Architecture Comparisons

There are many criteria that may be used to compare various parallel algorithms and architectures. The efficiency curves introduced in the previous section are one such mechanism which allow computational efficiency of each of a number of relevant algorithms to be compared across a range of machines. We intend to repeat the measurements described in this section for other parallel computers to provide a basis for comparison between machines. We have already measured the curves for two different algorithms, as seen in Figure 21 and Figure 22 and those graphs allow immediate comparison of the algorithms across a wide range of problem characteristics (*NFLOPS* and grid sizes). In fact the graphs are applicable to just about any situation where an areal sweep is performed over a rectangular grid and where only nearest-neighbor connectivity is involved between points. As an example, Figure 22 is directly applicable to the Areal Multigrid Relaxation discussed in section 12 below - one simply computes the number of floating point operations per grid-point in the relaxation, and can then read off the expected efficiency on different grid-sizes from the graph.

11. ELLIPTIC EQUATIONS AND SOLUTION METHODS

The basic elliptic equation we have studied is of the form:

$$\nabla \cdot (-\vec{K} \, \nabla P) \, (x,y) = F \, (x,y)$$

Here K has discontinuities, possibly of order a thousand or more, across a given set of curves and in typical applications may represent a fluid density, permeability or dielectric constant. The quantity of most interest is not the pressure or potential P but the velocity or flux $-\vec{K} \, \nabla P$. The right hand side F may contain arbitrary point and line sources. Boundary conditions may be Dirichlet, Neumann or mixed. We also allow a parabolic term on the left, provided an implicit time discretization is used. In the latter case, the following discussion will apply to the individual time steps

Discontinuities of coefficients imply discontinuities in the solution gradient Discretization of the equation on a rectangular grid leads to bad pressure and velocity solutions at the front due to such discontinuities. For this reason it is essential to adapt the grids locally In the resulting grids, discontinuities lie only along edges of triangles. The cost of grid generation is negligible compared to equation solution. We allow the use of curved edged triangles (isoparametrics) to provide high order boundary fitting. In general, our grids consist of unions of rectangles and triangles, with triangles primarily used to fit boundaries and interior interfaces For details of the grid construction methods used, we refer to our papers [19,20,21]

We have used finite element methods to discretize these equations, though similar studies could be applied to finite difference methods To provide sufficient accuracy, we allow high order elements up to cubics (on triangles) or bicubics (on rectangles) We have discussed the solution of singular elliptic equations by these techniques in [19,21] and the use of parallelism in the context of a tree of refinement grids elsewhere.[6] We have also previously described the solution of such systems on a parallel machine with global shared memory [4,6] We discuss here the implementation of fast solution methods on hypercube parallel processors. We describe the implementation of the *Conjugate Gradient Method* and an *FFT-based Fast Poisson Solver* which we have used as a preconditioner for the Conjugate Gradient Method We have also implemented a *Full Multi-grid Method* based on a five-point operator discretization of the equations. Using either of these methods, the solution cost in total operations performed is essentially proportional to the number of unknowns, while at the same time allowing near optimal use of parallelism.

Execution time is strongly dominated by the solution phase and it is the parallelization effort for this which we describe in the following sections We point out however that the finite element construction is also easily parallelized: the construction of the element matrices can proceed essentially independently on each element, fully utilizing available parallelism. Vectorizing such code is difficult, because the operations to be performed at individual finite element nodes are complicated, and because the elements are neither regular nor homogeneous (rectangles and triangles inter-mixed). Thus we will focus on parallel solution of the resulting equations and refer to our papers for details of the numerical analysis and of the discretization approach.[19,20,21,22,23]

12. PARALLEL MULTIGRID

We have developed a parallel multigrid algorithm and tested it on both the Caltech and Intel Hypercubes. Multigrid Methods are near-optimal for a wide range of computations on serial or vector processors Our test problem was a Poisson equation with Dirichlet boundary conditions on a square, discretized with a 5-point finite difference operator, but the implementation is considerably more general. A multigrid algorithm for the Poisson equation in 3 dimensions on the Caltech Hypercube has been developed independently by Clemens Thole.[24]

The basic multigrid idea[23,25,26,27] involves two aspects - the use of relaxation methods to dampen high-frequency errors and the use of multiple grids to allow low-frequencies to be relaxed inexpensively. A simple *Two-grid Iteration* involves a number of relaxations on a fine grid to reduce high-frequency errors in an initial guess of the solution, a projection of remaining errors to a coarser grid where they are solved for exactly, and then an interpolation back to the fine grid and addition to the solution there. This solution would now be exact but for errors introduced by the projection and interpolation processes. The solution is then improved by repeating the procedure.

The Two-grid Iteration is converted to the *Multigrid Iteration (MGI)* by recursively applying the 2-grid iteration in place of the exact solution on the coarse grid The number of times that the coarse grid iteration is repeated before returning to the fine grid is important for convergence rates - typical values used are once, known as *V-cycles*, or twice, known as *W-cycles*

Improved convergence can be obtained by choice of a good initial guess for the solution A simple strategy would be to solve the equations on a coarse grid using the Multigrid Iteration, and interpolate the solution to the fine grid as the initial guess for the Multigrid Iteration there Recursively applying this idea leads to the *Full Multigrid Iteration (FMG)* which performs a sequence of Full Multigrid solutions on increasingly finer grids, using the solution of each as an initial guess for the next.

Our studies are concerned with the case where there are many fine grid points per processor and we will assume this to be the case throughout the exposition. The algorithms presented in this section are valid for any architecture in which the processors can be arranged as a two dimensional periodic or non periodic array, depending on the nature of the boundary conditions of the elliptic problem We have developed both a Parallel Red-Black Gauss-Seidel relaxation operator and a Parallel Alternating Line Relaxation algorithm. The latter relaxation is important for problems which have variable coefficients, and is implemented using the parallel transpose operation.

Computational efficiency of the multigrid relaxation decreases with coarser grids because the perimeter to area ratio becomes less favorable as the number of grid-points is reduced. Consequently we find that V-cycles are somewhat more favored than W-cycles on a hypercube compared to a serial computer. Figure 23 shows a comparison of V and W-cycle speedup curves for the same sized grid which clearly demonstrates the superior computational efficiency of the V-cycle with respect to communication overhead. Similarly it is advantageous to terminate grid sub-division at a somewhat finer grid than one might use on a serial machine Note that in Figure 23 the curves begin only at speedups of 2 or 4 We were unable to solve the problem with that grid-size on only one processor due to memory limitations If the smallest number of processors that could be utilized was p then we arbitrarily assigned speedup p to the p-processor time

12.1. Distributed Grids for Multigrid

Having discretized the PDE using finite differences, the resulting equations on any grid level involve the solution or error on a rectangular grid of points. Parallelizing multigrid amounts to distributing these grids over the processors and implementing the communications needed as a result between processors. We accomplish the distribution by representing the solution (or error) on each grid level as a distributed matrix of type AREAL, see sections 7 1 and 7 3. In particular the hypercube is organized as a fixed rectangular mesh of $P = p_x p_y$ processors for all grid levels, using the matrix library call:

$$areal\_parameters(p_x, p_y, 0, 0, 1),$$

which provides a boundary of width 1 surrounding all sides of each subgrid allocated in subsequent allocate_matrix() calls for type AREAL, as well as non-periodic boundary conditions We will use the term *extended subgrid* to denote such a subgrid with boundary, and will refer to the non-boundary

- 94 -

points of the subgrid as its *interior* We assume that the grids to be used are of dimensions $2^{l}p_x \times 2^{l}p_y$ for some l_x, l_y With this choice, the interior of a level l subgrid assigned to a processor may be identified with a subset of the extended level $l+1$ subgrid assigned to that processor. Similarly the interior of the level $l+1$ subgrid in a processor is contained in the extended subgrid of level l.

12.2. Inter-process Communication

Along the boundary of the subgrids at any level some communication will be required by the multigrid relaxation operation. For example, a relaxation at boundary points of a sub-region will require the boundary data of the adjoining sub-region. The *shuffle*() operation from the matrix library, also described in section 7.3, provides exactly the required interchange of boundary data By calling *shuffle*() one can assure that each boundary edge at the start of any operation (relaxation or inter-grid transfer) holds a copy of the closest interior edge of the adjacent subgrid. For a 5-point operator, a boundary width of 1 suffices. The only possible problem here is the updating of corner nodes which *a priori* have to be copied from a non-neighboring processor. These are not actually needed in a five-point problem, but in more general situations are accommodated by adding an extra corner node shuffle. The *shuffle* operation is thus seen to be the basic communication routine necessary for multigrid to work on distributed grids

Inter-grid transfers between multigrid levels (injection and interpolation) are entirely local to each processor because of the containment relationships between the grids presented in the previous section. Thus no communication is involved in these steps. However before any updated grid values are used it is essential to perform a *shuffle*() operation to preserve the integrity of the duplicate boundary data. Thus we use the *shuffle*() operation not only before each relaxation but also before every inter-grid transfer.

It is in fact possible to choose a slightly different areal distribution of a matrix which avoids most of the shuffle operations at the cost of a small amount of memory. In this representation the interior edges of subgrids are actually duplicated in neighboring processors, as well as having a further boundary edge as above With this choice, the interior of the level $l+1$

subgrid contains all points of the interior of the level l subgrid, obviating any need to perform shuffles after inter-grid transfers. In this case a *shuffle* is required only before every relaxation, resulting in a substantial decrease in communication.

To control termination of iteration in adaptive multigrid algorithms, norms of error values need to be calculated These are evaluated by computing a local inner product in each processor, storing it in the local component of a distributed vector of length P, and then calling *sum_vector*() on this distributed vector, which computes the squared norm and distributes it to the processors - see the discussion of the semantics of *inner_product*() in section 6 5. If the alternate areal representation of the previous paragraph is used, it is important to weight interior edge points with a factor 5 and interior corner points with a factor .25 in computing local norms since these points are duplicated as interior points in neighboring subgrids.

12.3. The Relaxation Operation.

The relaxation operation is straightforward to implement in each of the subregions Standard Gauss-Seidel relaxation is discarded on numerical grounds of slower convergence. It introduces artificial discontinuities along the boundaries of the subregions as the parallel version uses the original values of the boundary duplicates. Note that in the limit of subregions consisting of only one grid point, this would reduce to a Jacobi relaxation Instead we utilize a Red-Black ordering of the grid points. This can be implemented such that the parallel algorithm will, after each iteration step, give the same result as the corresponding serial algorithm. The procedure consists of shuffling, relaxing the red points, shuffling again and then relaxing the black points.

12.4. Intergrid Transfers

The error (the residual equation to be precise) is projected to the coarse grid using half-injection - the errors at the black grid points are identically zero after the previous relaxation step. For this step as discussed previously, it is necessary that the extended fine grid subregions contain the non-extended coarse grid regions.

Having solved the error equation on the coarse grid the solution on the fine grid has to be updated by addition of a suitable interpolation of the computed coarse grid error. We have used linear interpolation at this point To do this in parallel without communication, the extended coarse grid subregion must contain the non-extended fine grid subregion We update the boundary duplicates in the fine grid subregions following the interpolation using a *shuffle* operation.

12.5. Results and Timings for Multigrid.

In Figure 23 we compare timings for 3-level V-cycle and W-cycle multigrid iterations with finest grid of dimension 64×64, and also a 4-level V-cycle iteration on a 128×128 grid These computations were performed on the Caltech hypercube with from 2 up to 32 nodes. The curves represent speedup attained with increasing number of processors (the 128 grid problem would not fit on 2 processors). We notice that the speedup for the W-cycle is worse than that for the V-cycle This is due to the fact that in the W-cycle algorithm, one spends a larger segment of the computation time on the coarser grids. In spite of this, it can be expected that for some equations it may be profitable to use the W-cycle instead of V-cycle because of its faster convergence rate. This issue however is too dependent on the particular type of equations solved to be discussed in this context. The V-cycle curve for the finer grid shows substantially improved performance compared to the coarser grid All of these curves correspond to relatively very coarse grids since the number of grid-points per processor is at most 512 on the finest grid

In Figure 24 we illustrate the effect of Area/Perimeter considerations on the Caltech Hypercube with 32 processors The hypercube was organized as logical rectangular grids of dimensions 32×1, 16×2 and 8×4, (see Figure 12) and we display the resulting computation time for a multigrid iteration on a 128×128 grid with 2 grid levels The improvement in timing as the subregions approach squares is obvious. The success of this demonstration is related to the fact that even for short messages communication costs on the Caltech Hypercube are strictly proportional to message length.

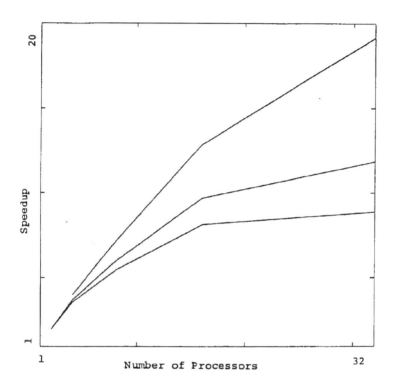

Fig. 23: The top curve is speedup versus Number of Processors for V cycle Multigrid on a 128 by 128 grid with 4 levels on the Caltech Hypercube. The two lower curves are for V and W cycle Multigrid on a 64 by 64 grid with 3 levels.

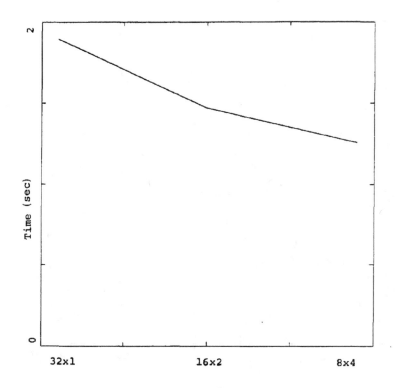

Fig. 24: Area-Perimeter considerations for Multigrid on the 32 node Caltech Hypercube. We compare times (in ms) for a V cycle multigrid iteration on a 128 by 128 grid with 2 levels, when the domain is split in 32 by 1, 16 by 2, 8 by 4 rectangles.

To illustrate that communication overhead decreases with an areal distribution as problem size increases, we compare in Figure 25 the scaled time of a multigrid iteration as a function of grid size. The scaled time is obtained by dividing the iteration time by a quantity proportional to the number of grid points and is therefore proportional to effective work per grid-point for multigrid *including* communication overhead. Specifically we compare the times for 16 iterations on a 64 by 64 grid, 4 iterations on a 128 by 128 grid and 1 iteration on a 256 by 256 grid on the 32 node Caltech Hypercube. Scaled time is over twice as small for the finest grid as for the coarsest.

Our multigrid results for the Intel iPSC, are cast in a somewhat different form We solved problems with varying grid sizes on a 128 node iPSC To compare the iteration times, we calculated the number of *useful* floating point operations (additions or multiplications, but not memory accesses) that were executed in each of the runs. We divided these by the time it took for an iteration to complete This defines an attained *floprate* for each run In Figure 26 we plot the attained floprate for V-cycles as a function of grid size, going from 4096 unknowns to over 1 6 million unknowns. We see that for the largest problems we attained about 3.75 megaflops.

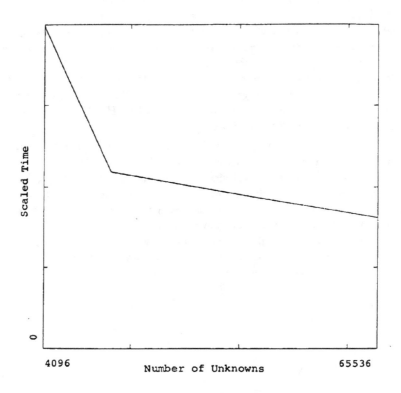

Fig. 25: Multigrid, 3 levels, V cycle on the Caltech Hypercube. We compare execution times of different sized problems by scaling the times by the number of floating point operations executed.

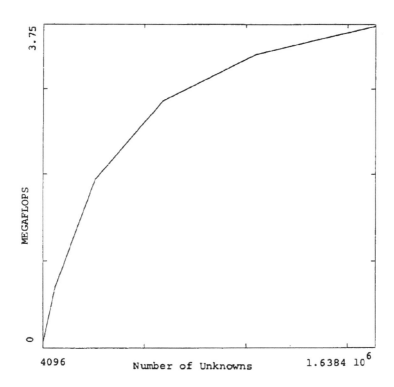

Fig. 26: Performance of 2 level V Cycle Multigrid Method on the 128 node iPSC. Number of Floating Point Operations executed per second as function of Problem Size.

13. PARALLEL CONJUGATE GRADIENT

Discretization of elliptic partial differential equations in two dimensions by finite element or finite difference methods leads to systems of equations with sparse coefficient matrices The fill-in of the matrix tends to follow diagonals and the bandwidth is about $dN^{1/2}$, where N is the dimension of the matrix and d is the degree of the finite elements used for the discretization. Furthermore typically only $O(1)$ diagonals have non-zero elements The Parallel Conjugate Gradient method we have developed on the hypercube, solves systems of equations with such coefficient matrix structures. This allows us to parallelize the solution of finite element discretizations of arbitrary and even variable degree with high efficiency.

13.1. The Serial Algorithm

The Preconditioned Conjugate Gradient Method[28, 29] finds the solution of the system of equations $Ax = f$, to a specified accuracy ϵ by performing the following iteration on the vector x, which has been appropriately initialized·

$$r = f - Ax$$
$$p = Br$$

loop

$$s = <r,Br>/<p,Ap>$$
$$r = r - s \cdot Ap$$
$$x = x + s\,p$$
$$rbr = <r,Br>$$
$$s = <r,Br>/old<r,Br>$$
$$p = Br + s\,p$$

until *converged*

Here B is an approximate inverse of A , which is assumed to be positive definite symmetric, and $<x,y>$ denotes the inner product of vectors x and y. The *preconditioning* operator B can be effective in improving substantially the convergence rate of the algorithm. In section 14, we will describe a parallel preconditioner for the conjugate gradient algorithm, concentrating in this section on the unaccelerated conjugate gradient method where B is just the identity operator.

13.2. Parallelism in Conjugate Gradient

We parallelize the algorithm by exploiting parallelism in every operation of the iteration. We have discussed previously how to parallelize each operation in the algorithm above (with the identity preconditioning) All of the vectors in the algorithm are allocated using the *allocate_vector*() library routine (see section 6) with type *SHIFT*. The matrix A is allocated using the *allocate_matrix*() library routine (see section 7) with type *DIAGONAL* and with the sparsity structure of the non-zero extended diagonals explicitly specified Each line in the algorithm above is expressed with the corresponding vector or matrix library routine. In particular the communication-intensive operation $p \rightarrow Ap$ is implemented directly using the *matrix_vector* library routine, so that all communication is performed by the optimized *shift_vector*() routine. It then suffices to compile the above program with the vector and matrix libraries introduced earlier. Note that the resulting program shows no dependence on a control processor. Each processor obtains the required inner products for convergence control using only a local subroutine call. All processors make the decision to stop iterating, and they will all do so simultaneously since they all receive the same value for the inner products.

13.3. Timing and Results for Conjugate Gradient

In Figure 27 we display the execution time per iteration step as a function of the number of processors used for the conjugate gradient method on a 64×64 grid on the Caltech Hypercube with from 2 up to 32 nodes In Figure 28 we compare the floprates attained in the execution of the conjugate

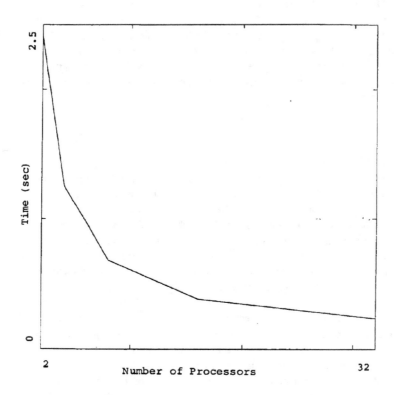

Fig. 27: Conjugate Gradient on a 64 by 64 grid. Time per Iteration step (in ms) versus Number of Processors used (Caltech Hypercube).

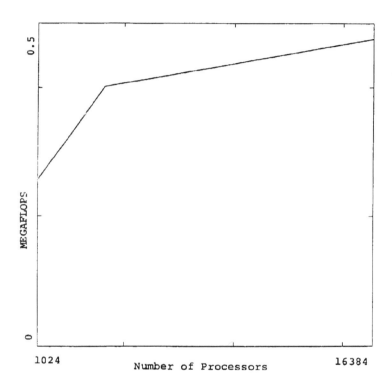

*Fig. 28: Performance of the Conjugate gradient method on the
32 node Caltech hypercube. We display the number of Floating
Point Operations executed per second for problems with 1024 up
16384 unknowns.*

gradient iteration on a 5D Caltech cube for different size grids. The floprates are determined by dividing the actual number of floating point operations executed by the the time to perform the iteration. We counted the number of multiplications in the algorithm to be about (*number of diagonals* + 5)*N* per iteration and about an equal number of floating point additions take place. These computations were performed with the non-optimized vector shift algorithm, and are for relatively coarse grids.

In Figure 29 we display the analogous graph for the 128 node iPSC for problems with 4096 up to 409600 unknowns. These computations were performed with the optimized shift algorithm and so are not directly comparable to the Caltech Hypercube results. Counting all useful floating point operations (multiplications and additions), the 128 node iPSC attained about 2 5 Megaflops in conjugate gradient iteration. The matrix used for the purpose of timings had 5 diagonals corresponding to linear finite elements.

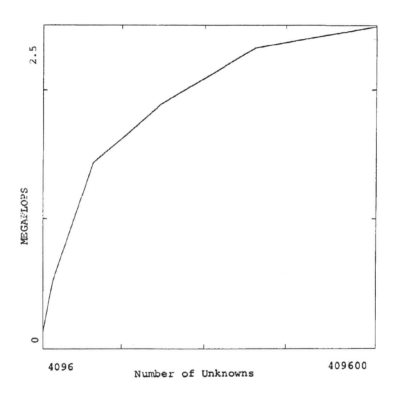

Fig. 29: Performance of the Conjugate Gradient Method on the 128 node iPSC. We display the number of Floating Point Operations executed per second for problems with 4096 up to 409,600 unknowns.

14. A PARALLEL FAST POISSON SOLVER

We have developed a parallel preconditioner for use in conjunction with the parallel conjugate gradient algorithm described in Section 13 The algorithm is based on the Fast Fourier transform and reduces the solution of the Poisson equation:

$$- \Delta\Phi(x,y) = F(x,y) ,$$

on a rectangle to Fourier transforms and tridiagonal solvers. This is a standard method frequently used on serial computers and the interest here is entirely in how to parallelize it   An attraction of our parallelization method is that neither the **FFT** nor the tridiagonal solvers need to be parallelized Assuming a 5-point discretization of the Poisson equation in a rectangle on an $m \times n$ grid, the solution may be obtained on a serial machine in three steps using the algorithm:

> *for col in* 1, . ,*m*   *fft*$(n, F(\cdot, col))$
>
> *for row in* 1, . . ,*n*   *tridiag*$(m, t(row, \cdot), F(row,))$
>
> *for col in* 1, . . ,*m*   *fft*$(n, F(\cdot, col))$

This algorithm returns the exact solution of the discrete Poisson equation in $O(nm \log n)$ operations on a serial computer. Here the matrix $F = F(,)$ is a discretization of the source function $F(x,y)$, while $F(row,)$ and $F(, col)$ denote the vectors corresponding to the indicated row or column of the matrix F The matrix $t(\cdot, \cdot)$ contains the coefficients of the tridiagonal systems, obtained analytically by Fourier transforming the five-point matrix The routine *fft*(n, v) computes a real (sine or cosine) fast Fourier transform of a vector v of length n. Similarly the routine *tridiag*(m, t, v) solves a tridiagonal system $Tu = v$, overwriting the m-component vector v with the solution u.

We have found that such a solver is a useful preconditioner even for higher-order finite elements on logically rectangular grids - one can use the Poisson solver for a rectangular grid corresponding in dimensionality to the nodal points of the finite elements, see McBryan.[19,23]

Vectorization of this algorithm involves considerable effort because it requires vectorizing the FFT routine which is non-trivial especially if the grid dimension n is more general than a power of 2. However parallelization is straight-forward The matrix F is stored as a distributed matrix of type *CONTIGUOUS_COLUMN*, while the matrix t is pre-computed and stored as a distributed matrix of type *CONTIGUOUS_ROW*, see section 7.1. Each *fft* and *tridiagonal* operation is assigned to an individual processor with each processor executing the normal serial code for that operation on one or more columns or rows respectively. This allows the full available parallelism to be applied effectively.

As we have noted in our previous Denelcor HEP studies, to complete the parallelization of this algorithm on a shared memory parallel computer it suffices to place a *barrier* between the first and second step, and again between the second and third step since it is necessary that all processors complete work on each step before any processor proceeds to the following step. There is a complication however on a hypercube stemming from the fact that the memory on the hypercube is not shared among the processors In the *fft* steps it is necessary to distribute the matrix F in contiguous-column form, so that the *fft()* operations are local to individual processors. Similarly in the tridiagonal step it is necessary to distribute the matrix F in contiguous-row form so that each tridiagonal solution involves only values from one processor To get from a row-distributed matrix to a column-distributed matrix requires a matrix transpose.

It follows that the above algorithm may be parallelized on a local-memory processor by distributing F in contiguous-column form and by inserting an extra *transpose(F)* step between the first and second step and between the second and third step. This along with the *barrier* described previously, are the only modifications required to the serial algorithm and in fact all communication for the Poisson solution is hidden in the two transpose operations On shared memory computers the *transpose* steps are not needed. However on machines with fast data caches there may be an advantage to using the transpose anyway, since it ensures that the *fft* and *tridiagonal* solutions work entirely on contiguous data. Another approach to parallelizing the fast Poisson solver, more cumbersome from a programming point of view, would be to adapt the Fourier transform algorithm to handle

distributed vectors This is non-trivial to perform unless the grid dimension
is a power of 2

15. Acknowledgements

We wish to thank Geoffrey Fox at Caltech for providing access to the
Caltech Hypercube and Shirley Enguehard at Caltech for much helpful
advice We also express our appreciation to Los Alamos National Laboratory
and to Intel Scientific Computers for access to their iPSC Hypercube systems,
and in particular to Ralph Brickner for his patience through many late night
and weekend system reboots and to Cleve Moler for providing documenta-
tion and other useful information.

References

1. O. McBryan and E. Van de Velde, "Parallel Algorithms for Elliptic
    Equations," *Proceedings of the 1984 ARO Novel Computing Environments
    Conference, Stanford University*, SIAM , to appear

2. O. McBryan, E. Van de Velde, and P. Vianna, "Parallel Algorithms for
    Elliptic and Parabolic Equations," *Proceedings of the Conference on
    Parallel Computations in Heat Transfer and Fluid Flows*, University of
    Maryland, November 1984

3   O. McBryan, "State of the Art of Multiprocessors in Scientific Computa-
    tion," *Proceedings of European Weather Center Conference on Multipro-
    cessors*, Reading, England, Dec 1984, to appear.

4. O. McBryan and E. Van de Velde, "Parallel Algorithms for Elliptic
    Equation Solution on the HEP Computer," *Proceedings of the First HEP
    Conference*, University of Oklahoma, March 1985.

5   O McBryan and E Van de Velde, "Elliptic Equation Algorithms on
    Parallel Computers," *Proceedings of the Conference on Parallel Comput-
    ers and Partial Differential Equations*, Commun in Applied Numerical

Methods, University of Texas, Austin, May 1986, to appear

6   O. McBryan and E. Van de Velde, "Parallel Algorithms for Elliptic Equations," *Commun. Pure and Appl. Math.*, Oct 1985

7   O. McBryan and E. Van de Velde, "Parallel Multigrid Implementations," *Proceedings of 2nd European Multigrid Conference, Cologne, Oct 1985*, to appear.

8. O. McBryan and E. Van de Velde, "Elliptic Equation Solution on Hypercube Multiprocessors," *Proceedings of the Oak Ridge meeting on Hypercube Architectures*, to appear, Aug 1985.

9. Charles L. Seitz, "The Cosmic Cube," *Communications of the ACM*, vol 28, No. 1, pp. 22-33, Jan. 1985.

10. G. C. Fox and S. W. Otto, "Algorithms for concurrent processors," *Phys Today*, vol. 37,5, pp. 50-59, May 1984.

11. John Salmon, "Binary Gray Codes and the Mapping of a Physical Lattice into a Hypercube," Caltech Concurrent Processor Report (CCP)Hm-51, 1983.

12. "Caltech/JPL Concurrent Computation Project Annual Report 1983-1984," Caltech Report, December 1984.

13 E. N Gilbert, "Gray Codes and Paths on the n-Cube," *Bell System Technical Journal*, vol 37, p 915, May 1958.

14. Martin Gardner, "Mathematical Games," *Scientific American*, p 106, August 1972.

15. T. Chan, Copper Mountain, Colorado, April 1985   Remark at 2nd Multigrid Conference

16. A. Chorin, "Random Choice Solution of Hyperbolic Systems," *J. Computational Phys* , vol. 22, pp. 517-533, 1976.

17. R. Richtmeyer and K. Morton , *Difference Methods for initial value problems*, Interscience Publishers , New York , 1952

18. J. Glimm , "Solutions in the Large for nonliner Hyperbolic Systems of Equations," *Comm. Pure Appl Math* , vol. 18, pp 697-715 , 1965

19. O. McBryan, "Computational Methods for Discontinuities in Fluids," *Lectures in Applied Mathematics*, vol 22, AMS, Providence, 1985

20. O. McBryan, "Elliptic and Hyperbolic Interface Refinement in Two Phase flow," in *Boundary and Interior Layers*, ed. J. J. H. Miller, Boole Press, Dublin , 1980.

21 O. McBryan , "Shock Tracking for 2d Flows," in *Computational and Assymptotic Methods for Boundary Layer Problems*, ed J.J. Miller, Boole Press, Dublin , 1982

22. J. Glimm and O McBryan, "A Computational Model for Interfaces," Courant Institute Preprint, May 1985.

23 O McBryan, "Fluids, Discontinuities and Renormalization Group methods," in *Mathematical Physics VII*, ed. Brittin, Gustafson and Wyss, pp 481-494, North-Holland Publishing Company, Amsterdam, 1984.

24 C Thole, "Experiments with Multigrid on the Caltech Hypercube," GMD Internal Report, October 1985.

25 A. Brandt, "Multi-level adaptive solutions to boundary-value problems," *Math Comp* , vol. 31, pp. 333-390, 1977.

26 W Hackbusch, "Convergence of multi-grid iterations applied to difference equations," *Math Comp* , vol 34, pp. 425-440, 1980

27 K Stuben and U Trottenberg, "On the construction of fast solvers for elliptic equations," *Computational Fluid Dynamics*, Rhode-Saint-Genese, 1982.

28 M. R Hestenes and E. Stiefel, "Methods of conjugate gradients for solving linear systems," *J. Res. Nat. Bur. Standards*, vol. 49, pp 409-436, 1952.

29 G. H. Golub and C. F. Van Loan, in *Matrix Computations*, John Hopkins Press, Baltimore, 1984.

6

www.ingramcontent.com/pod-product-compliance
Lightning Source LLC
LaVergne TN
LVHW012202040326
832903LV00003B/66